Andreas Prohl · Yanqing Wang

AF576138

Numerical Methods for Optimal Control Problems with SPDEs

Andreas Prohl
Mathematisches Institut
Eberhard Karls University of Tübingen
Tübingen, Germany

Yanqing Wang
School of Mathematics and Statistics
Southwest University
Chongqing, China

ISSN 2731-7595 ISSN 2731-7609 (electronic)
SpringerBriefs on PDEs and Data Science
ISBN 978-981-95-4468-4 ISBN 978-981-95-4469-1 (eBook)
https://doi.org/10.1007/978-981-95-4469-1

© The Editor(s) (if applicable) and The Author(s), under exclusive license to Springer Nature Singapore Pte Ltd. 2026

This work is subject to copyright. All rights are solely and exclusively licensed by the Publisher, whether the whole or part of the material is concerned, specifically the rights of translation, reprinting, reuse of illustrations, recitation, broadcasting, reproduction on microfilms or in any other physical way, and transmission or information storage and retrieval, electronic adaptation, computer software, or by similar or dissimilar methodology now known or hereafter developed.
The use of general descriptive names, registered names, trademarks, service marks, etc. in this publication does not imply, even in the absence of a specific statement, that such names are exempt from the relevant protective laws and regulations and therefore free for general use.
The publisher, the authors and the editors are safe to assume that the advice and information in this book are believed to be true and accurate at the date of publication. Neither the publisher nor the authors or the editors give a warranty, expressed or implied, with respect to the material contained herein or for any errors or omissions that may have been made. The publisher remains neutral with regard to jurisdictional claims in published maps and institutional affiliations.

This Springer imprint is published by the registered company Springer Nature Singapore Pte Ltd.
The registered company address is: 152 Beach Road, #21-01/04 Gateway East, Singapore 189721, Singapore

If disposing of this product, please recycle the paper.

SpringerBriefs on PDEs and Data Science

Editor-in-Chief

Enrique Zuazua, Department of Mathematics, University of Erlangen-Nuremberg, Erlangen, Germany

Series Editors

Paola Antonietti, Dipartimento di Matematica, MOX—Politecnico di Milano, Milano, Italy

Irene Fonseca, Department of Mathematical Sciences, Carnegie Mellon University, Pittsburgh, USA

Franca Hoffmann, Hausdorff Center for Mathematics, University of Bonn, Bonn, Germany

Shi Jin, Institute of Natural Sciences, Shanghai Jiao Tong University, Shanghai, China

Juan J. Manfredi, Department of Mathematics, University Pittsburgh, Pittsburgh, USA

Emmanuel Trélat, CNRS, Laboratoire Jacques-Louis Lions, Sorbonne University, Paris, France

Xu Zhang, School of Mathematics, Sichuan University, Chengdu, China

SpringerBriefs on PDEs and Data Science targets contributions that will impact the understanding of partial differential equations (PDEs), and the emerging research of the mathematical treatment of data science.

The series will accept high-quality original research and survey manuscripts covering a broad range of topics including analytical methods for PDEs, numerical and algorithmic developments, control, optimization, calculus of variations, optimal design, data driven modelling, and machine learning. Submissions addressing relevant contemporary applications such as industrial processes, signal and image processing, mathematical biology, materials science, and computer vision will also be considered.

The series is the continuation of a former editorial cooperation with BCAM, which resulted in the publication of 28 titles as listed here: https://www.springer.com/gp/mathematics/bcam-springerbriefs

Preface

This book is on the construction and convergence analysis of implementable algorithms to approximate the optimal control of a stochastic linear quadratic optimal control problem (SLQ problem, for short) subject to a stochastic PDE.

If compared to finite dimensional stochastic control theory, the increased complexity due to high-dimensionality requires new numerical concepts to approximate SLQ problems; likewise, well-established discretization and numerical optimization strategies from infinite dimensional deterministic control theory need fundamental changes to properly address the optimality system, where to approximate the solution of a backward stochastic PDE is conceptually new.

The linear-quadratic structure of SLQ problems allows two equivalent analytical approaches to characterize its minimum: ‘open loop’ is based on Pontryagin’s maximum principle, and ‘closed loop’ utilizes the stochastic Riccati equation in combination with the feedback control law. We will discuss why, in general, complexities of related numerical schemes differ drastically and when which direction should be given preference from an algorithmic viewpoint.

This book is a considerable extension of the results in [1–3], which we obtained over the last six years—by using tools from numerical stochastic analysis, statistics, as well as simulation. Obviously, many challenging problems still remain in the exciting interdisciplinary field of ‘numerical control theory for stochastic PDEs’, and we hope that this work will stimulate further research.

This work is partly supported by the National Natural Science Foundation of China under grant 11801467.

Tübingen, Germany — Andreas Prohl
Chongqing, China — Yanqing Wang
January 2024

References

1. A. Prohl, Y. Wang, Strong rates of convergence for a space-time discretization of the backward stochastic heat equation, and of a linear-quadratic control problem for the stochastic heat equation. ESAIM Control Optim. Calc. Var. **27**, Paper No. 54, 30 (2021)
2. A. Prohl, Y. Wang, Strong error estimates for a space-time discretization of the linearquadratic control problem with the stochastic heat equation with linear noise. IMA J. Numer. Anal. **42**, 3386–3429 (2022)
3. A. Prohl, Y. Wang, Convergence with rates for a Riccati-based discretization of SLQ problems with SPDEs. IMA J. Numer. Anal. **44**, 3393–3434 (2024)

Competing Interests The authors have no competing interests to declare that are relevant to the content of this manuscript.

Contents

Chapter 1
Introduction

Let D be a bounded domain with smooth boundary or a convex polyhedral domain in $\mathbb{R}^d, d = 1, 2, 3$, and $T > 0$. The main goal in this book is to *numerically approximate* the ($\mathbb{F}$-adapted) optimal control process $U^*(\cdot)$ that minimizes the *quadratic* cost functional ($\alpha \geq 0$)

$$\mathcal{J}\big(U(\cdot)\big) = \frac{1}{2}\mathbb{E}\Big[\int_0^T \big[\|X(t)\|^2 + \|U(t)\|^2\big]\,\mathrm{d}t + \alpha\|X(T)\|^2\Big] \tag{1.1}$$

subject to the *linear* stochastic partial differential equation (SPDE), with $\beta \in \mathbb{R}$,

$$\begin{cases} \mathrm{d}X(t) = \big[\Delta X(t) + U(t)\big]\mathrm{d}t + \big[\beta X(t) + \sigma(t)\big]\mathrm{d}W(t) \qquad t \in (0, T]\,, \\ X(0) = x \in \mathbb{H}_0^1 \cap \mathbb{H}^2\,, \end{cases} \tag{1.2}$$

which is supplemented with a homogeneous Dirichlet boundary condition. The concise formulation of the problem includes a given complete filtered probability space $(\Omega, \mathcal{F}, \mathbb{F}, \mathbb{P})$ on which a $\mathbb{R}$-valued standard Brownian motion W is defined, and will be referred to as
Problem (SLQ): 'minimize (1.1) subject to (1.2)'.

In this book, Problem **(SLQ)** serves as a prototypic formulation, for which numerical methods and their error analysis will be developed. Most of the results can be easily extended to more general stochastic optimal control problems. For example, the cost functional and the controlled equation could be taken as ($\alpha \geq 0\,, \delta > 0$)

$$\mathcal{J}\big(U(\cdot)\big)=\frac{1}{2}\mathbb{E}\Big[\int_0^T \big[\|X(t)-\widetilde{X}(t)\|^2+\delta\|U(t)\|^2\big]\,\mathrm{d}t+\alpha\|X(T)-\widetilde{X}(T)\|^2\Big],$$

where $\widetilde{X}(\cdot) : [0, T] \to \mathbb{L}^2$ is a given function, and ($m \in \mathbb{N}$)

© The Author(s), under exclusive license to Springer Nature Singapore Pte Ltd. 2026
A. Prohl and Y. Wang, *Numerical Methods for Optimal Control Problems with SPDEs*, SpringerBriefs on PDEs and Data Science,
https://doi.org/10.1007/978-981-95-4469-1_1

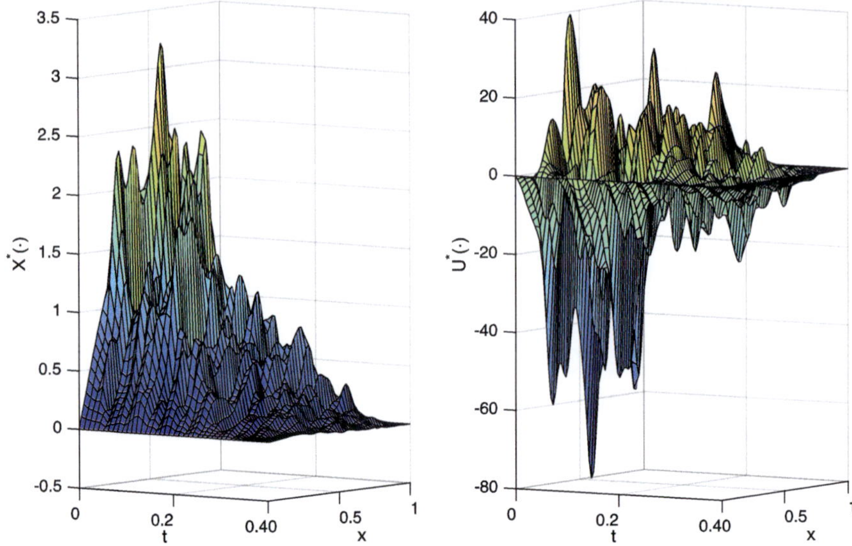

Fig. 1.1 Example 1.1: one simulated trajectory for the optimal state $X^*(\cdot)$ (left) and the optimal control $U^*(\cdot)$ (right)

$$\begin{cases} \mathrm{d}X(t) = \big[-\mathfrak{L}X(t) + \chi_{D_0} U(t)\big]\,\mathrm{d}t + \sum_{i=1}^{m} \big[\beta_i(t)X(t) + \sigma_i(t)\big]\,\mathrm{d}W_i(t) \quad t \in (0, T], \\ X(0) = x \in \mathbb{H}_0^1 \cap \mathbb{H}^2, \end{cases}$$

which is supplemented with homogeneous Dirichlet or Neumann boundary conditions. Here χ_{D_0} is the indicator function of the open subset D_0 of D, $W = (W_1, W_2, \cdots, W_m)$ is a $\mathbb{R}^m$-valued Brownian motion, and $\mathfrak{L}$ is the general strongly elliptic operator

$$\mathfrak{L}X = -\sum_{i,j=1}^{d} \frac{\partial}{\partial x_i}\Big(a_{ij}(\cdot)\frac{\partial X}{\partial x_j}\Big) + \sum_{j=1}^{d} b_j(\cdot)\frac{\partial X}{\partial x_j} + c(\cdot)X,$$

with the given functions $a_{ij}(\cdot)$, $b_j(\cdot)$, $c(\cdot)$.

Example 1.1 Let $D = (0, 1) \subset \mathbb{R}$, $T = 0.4$, and the system resp. the cost functional of Problem **(SLQ)** be

$$\mathrm{d}X(t) = \big[0.15\Delta X(t) + U(t)\big]\mathrm{d}t + \sum_{i=1}^{20} \beta_i X(t)\,\mathrm{d}W_i(t) \qquad t \in (0, T],$$

together with a homogeneous Dirichlet boundary condition, resp.

$$\mathcal{J}\big(U(\cdot)\big) = \frac{1}{2}\mathbb{E}\Big[\int_0^T \|X(t) - \widetilde{X}(t)\|^2 + \frac{1}{2000}\|U(t)\|^2\,\mathrm{d}t\Big],$$

where $X(0,\cdot) = 2\sin(\pi\cdot)$, $\beta_i(\cdot) = 2\sin(i\pi\cdot)$, $\widetilde{X}(t,\cdot) = \frac{2(T-t)}{T}\sin(\pi\cdot)$. Figure 1.1 displays a single path for an approximation of the optimal pair $\big(X^*(\cdot), U^*(\cdot)\big)$.

While the existence and uniqueness of a strong solution—which is the tuple $\big(X^*(\cdot), U^*(\cdot)\big)$ that minimizes Problem **(SLQ)**—is well-known by now, its explicit construction is usually not possible, and thus requires a numerical approximation: in fact, to construct an (*implementable*) *spatio-temporal discretization scheme* of a stochastic linear quadratic optimal control problem (SLQ problem, for short), and to verify the convergence of its solution is a very recent field in mathematical control theory, and the subject of this book, which combines tools from stochastic analysis, numerical optimization, and mathematical statistics.

Let us motivate the main conceptional difference in open-loop solvability between the numerics of the deterministic ($\beta = 0, \sigma(\cdot) = 0$) vs. the stochastic ($\beta \neq 0$ or $\sigma(\cdot) \neq 0$) control problem (1.1)–(1.2), which is the *adjoint equation* in the optimality conditions. The optimality conditions will be needed to formulate a gradient descent method, by which an implementable iterative algorithm can be proposed. We start with the deterministic case, for which different optimally convergent discretization methods are available.

(a) Let $\beta = 0$ and $\sigma(\cdot) = 0$. Then (1.2) is a *deterministic* PDE, and the overall problem is the *deterministic linear quadratic* optimal control problem (LQ problem, for short). The corresponding optimality system, which admits a unique solution $\big(X^*(\cdot), Y(\cdot), U^*(\cdot)\big)$, is given by

$$\begin{cases} \mathrm{d}X^*(t) = \big[\Delta X^*(t) + U^*(t)\big]\mathrm{d}t & t \in (0,T]\,, \quad (1.3a)\\ \mathrm{d}Y(t) = \big[-\Delta Y(t) + X^*(t)\big]\mathrm{d}t & t \in [0,T)\,, \quad (1.3b)\\ X^*(0) = x\,, \qquad Y(T) = -\alpha X^*(T)\,, & \quad (1.3c)\end{cases}$$

along with the optimality condition

$$U^*(t) - Y(t) = 0 \qquad \forall t \in [0,T]\,. \tag{1.4}$$

The deterministic PDEs in (1.3) are here written in the integral form to facilitate the direct comparison with the stochastic setting in (b) below. Clearly, (1.4) may be inserted into (1.3b), and the *adjoint equation* is a PDE running backward in time; hence, its complexity is of the same kind as the forward PDE (1.3a). Efficient and fast numerical methods that are based on the spatio-temporal discretization of this optimality system are now available; in addition, convergence with rates in terms of time and space discretization parameters may be shown, which is based on (discrete) stability properties and space-time regularity of the solution $\big(X^*(\cdot), U^*(\cdot)\big)$; see e.g. [1].

(b) Let $\beta \neq 0$ or $\sigma(\cdot) \neq 0$. The optimality condition for $\big(X^*(\cdot), U^*(\cdot)\big)$ constitutes a *coupled forward-backward SPDE system* (FBSPDE, for short) with a unique solution $\big(X^*(\cdot), Y(\cdot), Z(\cdot), U^*(\cdot)\big)$ (see e.g. [2]),

$$\begin{cases} \mathrm{d}X^*(t) = \big[\Delta X^*(t) + U^*(t)\big]\,\mathrm{d}t + \big[\beta X^*(t) + \sigma(t)\big]\,\mathrm{d}W(t) & t \in (0, T]\,, \quad (1.5a) \\ \mathrm{d}Y(t) = \big[-\Delta Y(t) - \beta Z(t) + X^*(t)\big]\,\mathrm{d}t + Z(t)\,\mathrm{d}W(t) & t \in [0, T)\,, \quad (1.5b) \\ X^*(0) = x\,, \qquad Y(T) = -\alpha X^*(T)\,, & \quad (1.5c) \end{cases}$$

together with the optimality condition

$$U^*(t) - Y(t) = 0 \qquad t \in [0, T]\,. \tag{1.6}$$

The adjoint equation (1.5b) is a *backward SPDE* (BSPDE, for short), whose solution now is a *tuple* $\big(Y(\cdot), Z(\cdot)\big)$, where the (additional) second component $Z(\cdot)$ plays its role to make sure that the first component $Y(\cdot)$ is $\mathbb{F}$-adapted; throughout the book, we assume that $\mathbb{F} = \{\mathcal{F}_t\}_{0 \le t \le T}$ is generated by the Brownian motion W. Its numerical approximation requires *significantly* larger computational resources if compared to PDE (1.3b), or SPDE (1.5a)—as will be motivated in the following paragraph. Next to a different algorithmic setup to approximate $\big(X^*(\cdot), U^*(\cdot)\big)$, it is also that the error analysis in Chap. 3 for a spatio-temporal discretization scheme of (1.5)–(1.6) differs profoundly from the deterministic case: a key step here is to bound temporal changes for (schemes of) $\big(X^*(\cdot), Y(\cdot), Z(\cdot), U^*(\cdot)\big)$ whose derivation requires tools from Malliavin calculus—where complications again enter which are caused by the driving Brownian motion.

If compared to the PDE (1.3b), we already hinted at the profoundly more complex endeavor to numerically simulate the stochastic adjoint equation (1.5b), which in fact is a BSPDE: after a finite element discretization of (1.5)–(1.6) in space (mesh width h), and in time (time step τ of a uniform mesh $I_\tau = \{t_n\}_{n=0}^N \subset [0, T]$) for $\beta = 0$ we have

$$\begin{cases} [\mathbb{1}_h - \tau\Delta_h]X^*_{h\tau}(t_{n+1}) = X^*_{h\tau}(t_n) + \tau U^*_{h\tau}(t_n) + \Pi_h\sigma(t_n)\Delta_{n+1}W\,, & (1.7a) \\ [\mathbb{1}_h - \tau\Delta_h]Y_{h\tau}(t_n) = \mathbb{E}^{t_n}\big[Y_{h\tau}(t_{n+1}) - \tau X^*_{h\tau}(t_{n+1})\big]\,, & (1.7b) \\ X^*_{h\tau}(0) = \Pi_h x\,, \qquad Y_{h\tau}(T) = -\alpha X^*_{h\tau}(T)\,, & (1.7c) \end{cases}$$

together with

$$U^*_{h\tau}(t_n) - Y_{h\tau}(t_n) = 0 \qquad 0 \le n \le N-1\,; \tag{1.8}$$

see Remark 3.5 in Sect. 3.3.3. Here, $\mathbb{1}_h$ denotes the identity operator on the finite element space, and $\Delta_{n+1}W = W(t_{n+1}) - W(t_n)$; the operator Δ_h denotes the discrete Laplacian, and Π_h is a projection that is related to finite elements; we refer to Chap. 3 for a detailed discussion.

The main effort now is to simulate the *conditional expectation* $\mathbb{E}^{t_n}[\cdot]$ in (1.7b); in fact, to accomplish this goal is a well-known task from regression analysis in *statistical learning*, where related estimators are available, including e.g. 'kNN-', partitioning-, or kernel estimators; see e.g. [3, Chap. 4]. These estimators perform well for problems where the dimension of the state space where $X^*_{h\tau}(\cdot)$ takes its values is *low*—such as stochastic control problems with *small* SDE-systems. In the present setting of Problem **(SLQ)** however, it is as *large* as the dimension of the finite element space $\mathbb{V}_h \subset \mathbb{H}^1_0$ that contains admissible states. Therefore, to efficiently cope with the related 'curse of dimensionality', we employ a 'data-dependent' partitioning estimator in Sect. 3.4.2—the construction of which was motivated by [4]—to approximate the conditional expectation in BSPDE (1.7b). However, the simulation is still costly, and requires to store large data sets to approximate the conditional expectation; and so is an iterative numerical solver based on 'gradient descent' for the whole system (1.7) which decouples simulations of the 'forward' and the 'backward part' per iteration. Note that this algorithmic complexity of (1.3)–(1.4) does not appear in the deterministic setting (1.1)–(1.2)—where to keep track of $\mathbb{F}$-adaptedness of iterates is *not* part of the problem. In Sect. 3.4.1, we therefore propose a conceptually different method that avoids the simulation of conditional expectations and hence cuts simulation times to only a fraction of the one where regression estimators are used—but this method is restricted to *additive* noise (i.e., $\beta = 0$). As a consequence, to judge the effectiveness of the 'open-loop' based numerics in the stochastic setting ($\beta \neq 0$ or $\sigma(\cdot) \neq 0$) requires to rethink established criteria from the deterministic case (such as efficiency, complexity,...), which e.g. often favor spatio-temporal discretization based on (1.5)–(1.6).

An advantage of the 'open-loop' based numerics in the stochastic setting is its applicability to even more general stochastic control problems than SLQ problems only. Consequently, the new numerical tools to accurately approximate the solution of BSPDEs in general, and in the context of (1.7)–(1.8) in particular, which are detailed in Chap. 3, are good to have. It is, however, that specific problems such as Problem **(SLQ)** additionally have a 'closed-loop' representation of the optimal control with the help of a feedback operator, which may be used to set up numerical schemes whose complexity shrinks drastically. The key idea here is to avoid BSPDEs, and use instead the (terminal backward) *stochastic Riccati equation*

$$\begin{cases} \mathcal{P}'(t) + \Delta\mathcal{P}(t) + \mathcal{P}(t)\Delta + \beta^2\mathcal{P}(t) + \mathbb{1} - \mathcal{P}^2(t) = 0 & t \in [0, T], \\ \mathcal{P}(T) = \alpha\mathbb{1}. \end{cases} \tag{1.9}$$

Here $\mathbb{1}$ denotes the identity operator on $\mathbb{L}^2$. Note that (1.9) is a deterministic equation with *operator-valued* solution, and the adjective 'stochastic' here only refers to its close relation to Problem **(SLQ)**, as is by now common in the literature; see e.g. [2, Chapter 13]. Once (an approximation of) $\mathcal{P}(\cdot)$ is available, another deterministic linear PDE needs to be solved

$$\begin{cases} \eta'(t) = -\Delta\eta(t) + \mathcal{P}(t)\eta(t) - \beta\mathcal{P}(t)\sigma(t) & t \in [0, T], \\ \eta(T) = 0, \end{cases} \tag{1.10}$$

where $\mathcal{P}(\cdot)$ now enters. Then, we may represent the optimal control $U^*(\cdot)$ of Problem **(SLQ)** with the help of the feedback law

$$U^*(t) = -\mathcal{P}(t)X^*(t) - \eta(t) \qquad t \in [0, T], \tag{1.11}$$

which may be inserted into (1.2). Hence, to solve (1.1)–(1.2) now only requires to solve spatio-temporal approximations of PDEs (1.9), (1.10), and of the SPDE (1.2). As a consequence, neither an iterative optimization algorithm ('gradient descent') is needed, nor the simulation of coupled FBSPDEs; it is, however, that a spatio-temporal discretization scheme of the stochastic Riccati equation (1.9) needs to be solved, which requires to solve for and store a finite sequence of large matrices $\{\mathcal{P}_n\}_{n=0}^N$. It turns out, however, that this is still cheaper (at least for $1d$ and $2d$ simulations) than to solve BSPDEs in the 'open-loop' approach—as comparative computational studies will show. Table 1.1 presents a comparative construction of the open-loop and closed-loop approaches.

We conclude this introduction with our own appreciation of the young research field: 'numerical control theory for SPDEs'—to which we wish to contribute with this book. It is (still a small) neighbor of the established field of 'numerical control theory for PDEs', whose continuing interest is nourished by practical applications. In this young area of numerical stochastic analysis, the last decade has seen only a few research papers, which may be due to the subject matter that combines tools from numerics, simulation, stochastic and deterministic PDEs. However, the practical relevance of this field is evident, and we expect a more rapid development of provably accurate, efficient algorithms in the near future; we conjecture this to happen also due

Table 1.1 Construction of different numerical schemes for Problem **(SLQ)**

Method	Open-loop approach	Closed-loop approach
Theory	Pontryagin's maximum principle: FBSPDE (1.5) with optimality condition (1.6)	Stochastic LQ theory: Riccati equation (1.9) and PDE (1.10)
Discretization strategies for Problem (SLQ)	**(a)** Spatial discretization by FEM for **(SLQ)**: a coupled FBSDE in finite dimensional space **(b)** Temporal discretization via the Euler method: a coupled stochastic difference equation **(c)** Decoupling by a gradient descent method	**(a)** Spatial discretization by FEM for **(SLQ)**: the Riccati equation in finite dimensional space **(b)** Temporal discretization via the Euler method: difference Riccati equation
Rates	$\mathcal{O}(h^2 + \tau^{1/2})$ by Theorems 3.6 and 3.8	$\mathcal{O}(h^2 \ln \frac{1}{h} + \tau^{1/2}[\lvert\beta\rvert + \tau^{1/2}])$ by Remark 4.5 and Theorem 4.7
	Improvement to $\mathcal{O}(h^2 + \tau)$ for additive noise by Theorem 3.6 for space and Theorem 4.7 for time	

to the related developments in theory, i.e.: ‘mathematical control theory for SPDEs’, which e.g. lead to the recent publication [2].

The book consists of two parts, each of which contains numerical schemes, their error analysis, and computational studies to illustrate their complexity:

(a) ‘open-loop’ based numerics in Chap. 3: we derive optimal rates of convergence for the spatio-temporal discretization of the optimality system (1.5)–(1.6), which is a coupled FBSPDE. In the second step, we decouple the computation of iterates in the course of a ‘gradient descent method’, which itself precedes an exact computation of conditional expectations ($\beta = 0$) or a regression analysis ($\beta \neq 0$). The mathematical tools needed to accomplish this goal are Malliavin calculus, (stochastic) Riccati equation analysis, (discrete) semigroup theory, and regression analysis. Simulations in `Python` will be reported that use conforming lowest-order finite element methods, and ‘data-dependent partitionings’ of the finite element space $\mathbb{V}_h$ to estimate conditional expectations.
(b) ‘closed-loop’ based numerics in Chap. 4: the tools needed here include optimal control analysis, the (stochastic) Riccati equation, and (discrete) semigroup theory. Simulations in `MATLAB` of (the stochastic) Riccati equation again use conforming lowest-order finite element methods.

These two schemes are also applicable to a broader class of SLQ problems. For instance, [5] employs the open-loop approach to numerically solve SLQ problems with Dirichlet boundary control, while [6] adopts a mixed approach—using the open-loop approach for spatial discretization and the closed-loop approach for temporal discretization—to address SLQ problems with distributed control in the diffusion term.

References

1. M. Hinze, R. Pinnau, M. Ulbrich, S. Ulbrich, Optimization with PDE constraints. *Mathematical Modelling: Theory and Applications*, vol. 23 (Springer, New York, 2009)
2. Q. Lü, X. Zhang, Mathematical control theory for stochastic partial differential equations. *Probability Theory and Stochastic Modelling*, vol. 101 (Springer, Cham, 2021)
3. L. Györfi, M. Kohler, A. Krzyżak, H. Walk, A distribution-free theory of nonparametric regression. *Springer Series in Statistics* (Springer, New York, 2002)
4. T. Dunst, A. Prohl, The forward-backward stochastic heat equation: numerical analysis and simulation. SIAM J. Sci. Comput. **38**, A2725–A2755 (2016)
5. A. Chaudhary, F. Merle, A. Prohl, Y. Wang, An efficient discretization to simulate the solution of linear-quadratic stochastic boundary control problem. IMA J. Numer. Anal., to appear
6. Y. Wang, Strong error estimates for the space-time discretization of a stochastic linear quadratic control problem with control in the diffusion. SIAM J. Control Optim. **63**, 3328–3355 (2025)

Chapter 2
Preliminaries

Section 2.1 introduces notations that will be used throughout the book, and collects tools to discretize Problem **(SLQ)**, in particular. Section 2.2 recalls bounds in strong norms for the solutions to linear SPDEs and BSPDEs, which are needed for the error analysis in Chaps. 3 and 4. The numerical schemes in these two chapters are based on two different concepts: 'open-loop approach' versus 'closed-loop approach', which are recalled in Sect. 2.3.

2.1 Notations, Assumptions, Methods and Tools

a. Spaces and operators. Let D be a bounded domain with smooth boundary or a convex polyhedral domain in $\mathbb{R}^d$, $d = 1, 2, 3$. By $\|\cdot\|$ resp. $(\cdot,\cdot)_{\mathbb{L}^2}$ we denote the norm resp. the scalar product in Lebesgue space $\mathbb{L}^2 \triangleq L^2(D)$. By $\|\cdot\|_{\mathbb{H}^{-1}}$, $\|\cdot\|_{\mathbb{H}_0^1}$, and $\|\cdot\|_{\mathbb{H}^2}$, we denote norms in Sobolev spaces $\mathbb{H}^{-1} \triangleq H^{-1}(D)$, $\mathbb{H}_0^1 \triangleq H_0^1(D)$, and $\mathbb{H}^2 \triangleq H^2(D)$ respectively. The duality pairing between $\mathbb{H}^{-1}$ and $\mathbb{H}_0^1$ is denoted by $\langle\cdot,\cdot\rangle$.

Fix $T > 0$, and let $(\Omega, \mathcal{F}, \mathbb{F}, \mathbb{P})$ be a complete filtered probability space, where $\mathbb{F} = \{\mathcal{F}_t\}_{t\in[0,T]}$ is the natural filtration generated by the $\mathbb{R}$-valued Brownian motion W, which is augmented by all the $\mathbb{P}$-null sets. Denote by $\mathbb{E}^t[\xi]$ the conditional expectation $\mathbb{E}[\xi \,|\, \mathcal{F}_t]$.

For a given Hilbert space $\big(\mathbb{H}, (\cdot,\cdot)_{\mathbb{H}}\big)$, we set

$$\mathbb{S}(\mathbb{H}) \triangleq \big\{\mathcal{P} \in \mathcal{L}(\mathbb{H}) \,|\, \mathcal{P} \text{ is symmetric}\big\},$$

and

$$\mathbb{S}_+(\mathbb{H}) \triangleq \big\{\mathcal{P} \in \mathbb{S}(\mathbb{H}) \,|\, (\mathcal{P}\eta, \eta)_{\mathbb{H}} \geq 0 \quad \forall\, \eta \in \mathbb{H}\big\}.$$

© The Author(s), under exclusive license to Springer Nature Singapore Pte Ltd. 2026
A. Prohl and Y. Wang, *Numerical Methods for Optimal Control Problems with SPDEs*, SpringerBriefs on PDEs and Data Science,
https://doi.org/10.1007/978-981-95-4469-1_2

For any $t \in [0, T]$, let $L^2_{\mathcal{F}_t}(\Omega; \mathbb{H}) \triangleq L^2(\Omega, \mathcal{F}_t, \mathbb{P}; \mathbb{H})$, and

$$
\begin{aligned}
L^2_{\mathbb{F}}(0, T; \mathbb{H}) &\triangleq \Big\{\psi : (0, T)\times\Omega \to \mathbb{H} \;\Big|\; \psi(\cdot) \text{ is } \mathbb{F}\text{-adapted} \\
&\qquad \text{and } \mathbb{E}\Big[\int_0^T \|\psi(t)\|^2_{\mathbb{H}}\, dt\Big] < \infty\Big\}, \\
L^2_{\mathbb{F}}\big(\Omega; C([0, T]; \mathbb{H})\big) &\triangleq \Big\{\psi \in L^2_{\mathbb{F}}(0, T; \mathbb{H}) \;\Big|\; \psi(\cdot) \text{ admits a continuous path} \\
&\qquad \text{and } \sup_{t\in[0,T]} \mathbb{E}\big[\|\psi(t)\|^2_{\mathbb{H}}\big] < \infty\Big\}, \\
C_{\mathbb{F}}\big([0, T]; L^2(\Omega; \mathbb{H})\big) &\triangleq \Big\{\psi \in L^2_{\mathbb{F}}(0, T; \mathbb{H}) \;\Big|\; \psi(\cdot) \text{ is } \mathbb{F}\text{-adapted} \\
&\qquad \text{and } \varphi(\cdot) : [0, T] \to L^2_{\mathcal{F}_T}(\Omega; \mathbb{H}) \text{ is continuous}\Big\}, \\
D_{\mathbb{F}}\big([0, T]; L^2(\Omega; \mathbb{H})\big) &\triangleq \Big\{\psi \in L^2_{\mathbb{F}}(0, T; \mathbb{H}) \;\Big|\; \psi(\cdot) \text{ is } \mathbb{F}\text{-adapted} \\
&\qquad \text{and } \varphi(\cdot) : [0, T] \to L^2_{\mathcal{F}_T}(\Omega; \mathbb{H}) \text{ is cadlag}\Big\}.
\end{aligned}
$$

For any $a\,, b \in \mathbb{R}$, set

$$a \vee b \triangleq \max\{a\,, b\}\,, \qquad a \wedge b \triangleq \min\{a\,, b\}\,. \tag{2.1}$$

For a sequence $\{a_j\}_{j\geq 0}$, we adopt the convention that $\sum_{j=k}^{l} a_j = 0$ and $\prod_{j=k}^{l} a_j = 1$ when $l < k$.

Below, let C be a generic positive constant which may depend on the data of the specific problems.

Let $\Delta : \mathbb{H}^1_0 \cap \mathbb{H}^2 \to \mathbb{L}^2$ be the Dirichlet realization of the Laplace operator. Naturally, Δ generates the analytic semigroup $E(\cdot) \equiv \{E(t) \mid t \geq 0\}$, with $E(t) = e^{t\Delta}$ for $t \geq 0$. Moreover, there exist an increasing sequence of eigenvalues $\{\lambda_n\}_{n\geq 1}$ and an orthonormal basis of eigenvectors $\{\varphi_n\}_{n\geq 1}$ in $\mathbb{L}^2$ such that $\Delta\varphi_n = -\lambda_n\varphi_n$ and

$$0 < \lambda_1 \leq \lambda_2 \leq \cdots (\to \infty)\,.$$

By the spectral decomposition for Δ, we can define fractional powers $(-\Delta)^\gamma$ for $\gamma \geq 0$ by

$$(-\Delta)^\gamma x = \sum_{n=1}^{\infty} \lambda_n^\gamma (x, \varphi_n)_{\mathbb{L}^2}\varphi_n\,,$$

where $x \in D\big((-\Delta)^\gamma\big) \triangleq \big\{x \in \mathbb{L}^2 \mid \sum_{n=1}^{\infty} \lambda_n^{2\gamma}(x, \varphi_n)^2_{\mathbb{L}^2} < \infty\big\}$. Also, we introduce $\dot{\mathbb{H}}^\gamma \triangleq D\big((-\Delta)^{\gamma/2}\big)$, and have $\|x\|_{\dot{\mathbb{H}}^\gamma} = \|(-\Delta)^{\gamma/2}x\|$ for any $x \in \dot{\mathbb{H}}^\gamma$. Obviously, $\|\cdot\| = \|\cdot\|_{\dot{\mathbb{H}}^0}$. Following [1], we know that

$$\dot{\mathbb{H}}^1 = \mathbb{H}^1_0\,, \qquad \dot{\mathbb{H}}^2 = \mathbb{H}^1_0 \cap \mathbb{H}^2\,.$$

For $\gamma > 0$, we consider the set

$$\dot{\mathbb{H}}^{-\gamma} \triangleq \Big\{ x = \sum_{n=1}^{\infty} x_n \varphi_n \,\Big|\, x_n \in \mathbb{R}, n = 1, 2, \cdots, \text{ such that } \|x\|^2_{\dot{\mathbb{H}}^{-\gamma}} = \sum_{n=1}^{\infty} \lambda_n^{-\gamma} x_n^2 < \infty \Big\} .$$

For any $x = \sum_{n=1}^{\infty} x_n \varphi_n \in \dot{\mathbb{H}}^{-\gamma}$, we also can define the fractional power of $-\Delta$ for negative exponents by

$$(-\Delta)^{-\gamma/2} x = \sum_{n=1}^{\infty} \lambda_n^{-\gamma/2} x_n \varphi_n .$$

By [2, Theorem B8], $\dot{\mathbb{H}}^{-\gamma}$ is a separable Hilbert space, which is isometrically isomorphic to $(\dot{\mathbb{H}}^{\gamma})'$.

The following properties of the semigroup $E(\cdot)$ will be utilized frequently throughout this work.

Lemma 2.1 *For the analytic semigroup $E(\cdot)$, there exists a constant C such that the following properties hold true:*

$$\begin{cases} \big\|(-\Delta)^{\gamma} E(t)\big\|_{\mathcal{L}(\mathbb{L}^2)} \le C t^{-\gamma} & \forall \gamma \ge 0, t > 0; & \text{(2.2a)} \\ \big\|(-\Delta)^{-\gamma}\big(E(t) - \mathbb{1}\big)\big\|_{\mathcal{L}(\mathbb{L}^2)} \le C t^{\gamma} & \forall \gamma \in [0, 1], t \ge 0; & \text{(2.2b)} \\ \|E(t) - E(s)\|_{\mathcal{L}(\mathbb{L}^2)} \le C (t-s)^{\gamma} s^{-\gamma} & \forall \gamma \in [0, 1], \ s, t > 0, s < t. & \text{(2.2c)} \end{cases}$$

The assertions (2.2a) and (2.2b) can be found in [3], and then (2.2c) can be derived directly.

In this book, we make the following assumptions on further data that appear in problem (1.1)–(1.2).
Assumption (A): Let $x \in \mathbb{H}_0^1 \cap \mathbb{H}^2$, and $\sigma \in C([0, T]; \mathbb{H}_0^1) \cap L^2(0, T; \mathbb{H}_0^1 \cap \mathbb{H}^2)$. There exists a positive constant $L_{\sigma,\alpha}$, for $\alpha = 1/2\,, 1$ such that

$$\|\sigma(t) - \sigma(s)\| \le L_{\sigma,\alpha} |t - s|^{\alpha} \qquad \forall t, s \in [0, T] .$$

b. Malliavin derivative. We use Malliavin derivatives of involved processes in BSPDE (2.18b) below to quantify temporal changes of $\big(X^*(\cdot), U^*(\cdot)\big)$ in Problem **(SLQ)**; see e.g. Lemma 2.7 in Sect. 2.3. Hence, we briefly recall the definition and key results of the Malliavin derivative of processes that will be needed below. For further details, we refer to [4].

We define $\mathbb{W}(\cdot) : L^2(0, T; \mathbb{R}) \to L^2_{\mathcal{F}_T}(\Omega; \mathbb{R})$ by

$$\mathbb{W}(g) = \int_0^T g(t)\,\mathrm{d}W(t) .$$

For $\ell \in \mathbb{N}$, we denote by $C_p^\infty(\mathbb{R}^\ell)$ the space of all smooth functions $f : \mathbb{R}^\ell \to \mathbb{R}$ such that $f(\cdot)$ and all of its partial derivatives have polynomial growth. Let Q be the set of $\mathbb{R}$-valued random variables of the form

$$F = f\big(\mathbb{W}(g_1), \mathbb{W}(g_2), \cdots, \mathbb{W}(g_\ell)\big)$$

for some $f(\cdot) \in C_p^\infty(\mathbb{R}^\ell)$, $\ell \in \mathbb{N}$, and $g_1(\cdot), \cdots, g_\ell(\cdot) \in L^2(0, T; \mathbb{R})$. For any $F \in Q$ we define its $\mathbb{R}$-valued Malliavin derivative process $DF \equiv \{D_\theta F \mid 0 \le \theta \le T\}$ via

$$D_\theta F = \sum_{i=1}^{\ell} \frac{\partial f}{\partial x_i}\big(\mathbb{W}(g_1), \mathbb{W}(g_2), \ldots, W(g_\ell)\big) g_i(\theta).$$

In general, we define the k-th iterated derivative of F by $D^k F = D(D^{k-1}F)$, for any $k \in \mathbb{N}$. Note that for any $\theta \in [0, T]$, $D_\theta F$ is $\mathcal{F}_T$-measurable; and if F is $\mathcal{F}_t$-measurable, then $D_\theta F = 0$ for any $\theta \in (t, T]$.

Now we can extend the derivative operator to $\mathbb{H}$-valued random variables, where $\big(\mathbb{H}, (\cdot, \cdot)_\mathbb{H}\big)$ is a real separable Hilbert space. For any $k \in \mathbb{N}$, and u in the set of $\mathbb{H}$-valued variables:

$$Q_\mathbb{H} = \Big\{u = \sum_{j=1}^{n} F_j \phi_j \,\Big|\, F_j \in Q,\ \phi_j \in \mathbb{H},\ n \in \mathbb{N}\Big\},$$

we can define the k-th iterated derivative of u by $D^k u = \sum_{j=1}^n D^k F_j \otimes \phi_j$. For $p \ge 1$, we define the norm $\|\cdot\|_{k,p}$ via

$$\|u\|_{k,p} \triangleq \Big(\mathbb{E}\big[\|u\|_\mathbb{H}^p\big] + \sum_{j=1}^{k} \mathbb{E}\big[\big\|D^j u\big\|^p_{\left(L^2(0,T;\mathbb{R})\right)^{\otimes j} \otimes \mathbb{H}}\big]\Big)^{1/p}.$$

Then $\mathbb{D}^{k,p}(\mathbb{H})$ is the completion of $Q_\mathbb{H}$ under the norm $\|\cdot\|_{k,p}$.

c. Discretization in space – the finite element method. We partition the bounded domain $D \subset \mathbb{R}^d$ via a regular triangulation $\mathcal{T}_h$ into elements K with maximum mesh size $h \triangleq \max\{\mathrm{diam}(K) \mid K \in \mathcal{T}_h\}$, and consider the space

$$\mathbb{V}_h \triangleq \big\{\phi \in \mathbb{H}_0^1 \,\big|\, \phi|_K \in \mathbb{P}_1(K) \quad \forall K \in \mathcal{T}_h\big\},$$

where $\mathbb{P}_1(K)$ denotes the space of polynomials of degree 1; see e.g. [5]. We define the discrete Laplacian $\Delta_h : \mathbb{V}_h \to \mathbb{V}_h$ by $(-\Delta_h \varphi_h, \phi_h)_{\mathbb{L}^2} = (\nabla \varphi_h, \nabla \phi_h)_{\mathbb{L}^2}$ for all $\varphi_h, \phi_h \in \mathbb{V}_h$, the generalized $\mathbb{L}^2$-projection $\Pi_h : \mathbb{H}^{-1} \to \mathbb{V}_h$ by $(\Pi_h x, \phi_h)_{\mathbb{L}^2} = \langle x, \phi_h\rangle$ for all $x \in \mathbb{H}^{-1}$, $\phi_h \in \mathbb{V}_h$, and the Ritz-projection $\mathcal{R}_h : \mathbb{H}_0^1 \to \mathbb{V}_h$ by $(\nabla[\mathcal{R}_h x - x], \nabla \phi_h)_{\mathbb{L}^2} = 0$ for all $x \in \mathbb{H}_0^1$, $\phi_h \in \mathbb{V}_h$. It is evident when $x \in \mathbb{L}^2$, that $\Pi_h x$ is the standard $\mathbb{L}^2$-projection; see e.g. [6]. Via definitions of Π_h, $\mathcal{R}_h$, it is easy to get that

$$\begin{cases} \|\Pi_h x\| \le \|x\| & \forall\, x \in \mathbb{L}^2\,, \\ \|\nabla \mathcal{R}_h x\| \le \|\nabla x\| & \forall\, x \in \mathbb{H}_0^1\,. \end{cases} \tag{2.3}$$

Throughout this work, we fix a constant $h_0 \in (0, 1)$, such that $h \in (0, h_0]$, and use the notation $C_{h_0} = \frac{\ln T}{\ln \frac{1}{h_0}}$.

Remark 2.1 Similar to Δ, the finite rank operator Δ_h also generates a semigroup $E_h(\cdot) \equiv \{E_h(t)\,|\,t \ge 0\}$, where $E_h(t) = e^{t\Delta_h}$. Besides, $E_h(\cdot)$ also enjoys the properties stated in Lemma 2.1. See e.g. [1]. Furthermore, there exist an increasing sequence of eigenvalues $\{\lambda_{h,n}\}_{n=1}^{\dim(\mathbb{V}_h)}$, and an orthonormal basis of eigenvectors $\{\varphi_{h,n}\}_{n=1}^{\dim(\mathbb{V}_h)}$ in $\mathbb{L}^2$ such that $\Delta_h \varphi_{h,n} = -\lambda_{h,n}\varphi_{h,n}$, where

$$0 < \lambda_{h,1} \le \lambda_{h,2} \le \cdots \le \lambda_{h,\dim(\mathbb{V}_h)}\,.$$

By the spectral decomposition for Δ_h, we can define fractional powers $(-\Delta_h)^\gamma$ for $\gamma \in \mathbb{R}$ by the following:

$$(-\Delta_h)^\gamma x = \sum_{n=1}^{\dim(\mathbb{V}_h)} \lambda_{h,n}^\gamma (x, \varphi_{h,n})_{\mathbb{L}^2} \varphi_{h,n}\,,$$

where $x \in D\big((-\Delta_h)^\gamma\big) \triangleq \big\{x \in \mathbb{V}_h \,\big|\, \sum_{n=1}^{\dim(\mathbb{V}_h)} \lambda_{h,n}^{2\gamma}(x, \varphi_{h,n})_{\mathbb{L}^2}^2 < \infty\big\}$. Also, we introduce $\dot{\mathbb{H}}_h^\gamma \triangleq D\big((-\Delta_h)^{\gamma/2}\big)$, and define $\|x\|_{\dot{\mathbb{H}}_h^\gamma} = \|(-\Delta_h)^{\gamma/2}x\|$ for any $x \in \dot{\mathbb{H}}_h^\gamma$. Obviously for any $x \in \mathbb{V}_h$, $\|x\|_{\dot{\mathbb{H}}_h^0} = \|x\|$ and $\|x\|_{\dot{\mathbb{H}}_h^1} = \|x\|_{\dot{\mathbb{H}}^1}$.

The following result lists some properties on Π_h and $\mathcal{R}_h$, which will be frequently applied.

Lemma 2.2 *There exists a constant C such that*

$$\begin{cases} \|x - \Pi_h x\| + h\|x - \Pi_h x\|_{\mathbb{H}_0^1} \le Ch^\gamma \|x\|_{\dot{\mathbb{H}}^\gamma} & \forall\, x \in \dot{\mathbb{H}}^\gamma\,, \gamma = 1, 2\,, & (2.4a) \\ \|x - \mathcal{R}_h x\| + h\|x - \mathcal{R}_h x\|_{\mathbb{H}_0^1} \le Ch^\gamma \|x\|_{\dot{\mathbb{H}}^\gamma} & \forall\, x \in \dot{\mathbb{H}}^\gamma\,, \gamma = 1, 2\,, & (2.4b) \\ \|x - \Pi_h x\|_{\dot{\mathbb{H}}^{-1}} \le Ch^2 \|x\|_{\mathbb{H}_0^1} & \forall\, x \in \mathbb{H}_0^1\,, & (2.4c) \\ \|\Delta_h \Pi_h x\| \le C\|x\|_{\dot{\mathbb{H}}^2} & \forall\, x \in \dot{\mathbb{H}}^2\,, & (2.4d) \\ \|(\Pi_h \Delta - \Delta_h \Pi_h)x\|_{\dot{\mathbb{H}}_h^{-\gamma}} \le Ch^\gamma \|x\|_{\dot{\mathbb{H}}^2} & \forall\, x \in \dot{\mathbb{H}}^2\,, \gamma = 1, 2\,. & (2.4e) \end{cases}$$

Proof **(1)** The first two estimates are standard; see e.g. [5, 6]. For the assertion (2.4c), we use $\dot{\mathbb{H}}^{-1} = \mathbb{H}^{-1} \triangleq \big(\mathbb{H}_0^1\big)^*$, and conclude as follows:

$$\|x - \Pi_h x\|_{\dot{\mathbb{H}}^{-1}} = \sup_{\varphi \in \mathbb{H}_0^1} \frac{(x - \Pi_h x, \varphi - \Pi_h \varphi)_{\mathbb{L}^2}}{\|\varphi\|_{\mathbb{H}_0^1}}$$

$$\leq Ch^2 \|x\|_{\mathbb{H}_0^1} \sup_{\varphi \in \mathbb{H}_0^1} \frac{\|\varphi\|_{\mathbb{H}_0^1}}{\|\varphi\|_{\mathbb{H}_0^1}} \leq Ch^2 \|x\|_{\mathbb{H}_0^1} .$$

(2) For any $x \in \mathbb{H}_0^1 \cap \mathbb{H}^2$ and $\varphi \in \mathbb{L}^2$, by using the identity

$$(-\Delta_h \mathcal{R}_h x, \Pi_h \varphi)_{\mathbb{L}^2} = (\nabla \mathcal{R}_h x, \nabla \Pi_h \varphi)_{\mathbb{L}^2} = (\nabla x, \nabla \Pi_h \varphi)_{\mathbb{L}^2} = (-\Delta x, \Pi_h \varphi)_{\mathbb{L}^2} , \tag{2.5}$$

which follows from the definition of $\mathcal{R}_h$ and integration by parts. Then relying on an inverse estimate (*cf.* [5]), assertions (2.4a)–(2.4b) and (2.3) on $\mathbb{L}^2$-stability of Π_h, we conclude that

$$\begin{aligned}
(\Delta_h \Pi_h x, \Pi_h \varphi)_{\mathbb{L}^2} &= (\Delta_h \mathcal{R}_h x, \Pi_h \varphi)_{\mathbb{L}^2} + (\Delta_h [\Pi_h - \mathcal{R}_h] x, \Pi_h \varphi)_{\mathbb{L}^2} \\
&= (\Delta x, \Pi_h \varphi)_{\mathbb{L}^2} - (\nabla [\Pi_h - \mathbb{1}] x, \nabla \Pi_h \varphi)_{\mathbb{L}^2} + (\nabla [\mathcal{R}_h - \mathbb{1}] x, \nabla \Pi_h \varphi)_{\mathbb{L}^2} \\
&\leq \|\Delta x\| \|\Pi_h \varphi\| + Ch \|x\|_{\dot{\mathbb{H}}^2} \frac{1}{h} \|\Pi_h \varphi\| \\
&\leq C \|x\|_{\dot{\mathbb{H}}^2} \|\varphi\| ,
\end{aligned}$$

which yields (2.4d).

(3) By (2.5), it follows that for any $x \in \dot{\mathbb{H}}^2$, $\Delta_h \mathcal{R}_h x = \Pi_h \Delta x$. Then

$$\begin{aligned}
(-\Delta_h)^{-1} (\Pi_h \Delta - \Delta_h \Pi_h) x &= (-\Delta_h)^{-1} (\Delta_h \mathcal{R}_h - \Delta_h \Pi_h) x \\
&= (\mathcal{R}_h - \mathbb{1}) x + (\mathbb{1} - \Pi_h) x ,
\end{aligned}$$

which, together with (2.4a)–(2.4b), leads to the assertion (2.4e) for $\gamma = 2$. The estimate for $\gamma = 1$ can be derived similarly. □

For errors committed by the finite element method, we define the error operator

$$G_h(\cdot) \triangleq E(\cdot) - E_h(\cdot) \Pi_h . \tag{2.6}$$

The following result will be frequently used in later chapters; see e.g. [7, Lemma 3.1].

Lemma 2.3 *There exists a constant C, such that the following estimate for the error operator $G_h(\cdot)$ holds: for any $h \in (0, 1]$, $t \in (0, T]$,*

$$\|G_h(t) x\| \leq Ch^{\gamma} t^{-(\gamma - \rho)/2} \|x\|_{\dot{\mathbb{H}}^\rho} \qquad \forall x \in \dot{\mathbb{H}}^\rho , 0 \leq \rho \leq \gamma \leq 2 . \tag{2.7}$$

d. Discretization in time—interpolation and discrete semigroup $E_{h,\tau}(\cdot)$. We denote by $I_\tau = \{t_n\}_{n=0}^N \subset [0, T]$ a uniform time mesh with step size $\tau \triangleq T/N$ and $t_n = n\tau$, and related Wiener increments by $\Delta_n W \triangleq W(t_n) - W(t_{n-1})$ for all $n = 1, \cdots, N$. For a given time mesh I_τ, we can define piecewise constant functions $\mu(\cdot)$, $\nu(\cdot)$, $\pi(\cdot)$ by

$$\mu(t) = t_{n+1}\,, \qquad \nu(t) = t_n\,, \qquad \pi(t) = n+1\,, \tag{2.8}$$

where $t \in [t_n, t_{n+1})$, $n = 0, 1, \cdots, N-1$; we will use the (piecewise constant) operator $\Pi_\tau : C([0,T];\mathbb{H}) \to D([0,T];\mathbb{H})$ defined by

$$\Pi_\tau f(t) = f(t_n) \qquad \forall\, t \in [t_n, t_{n+1}) \qquad n = 0, 1, \cdots, N-1\,. \tag{2.9}$$

Throughout this work, we fix some constant $\tau_0 \in (0,1)$ such that $\tau \in (0, \tau_0]$ and introduce a related constant $C_{\tau_0} = \frac{\ln T}{\ln \frac{1}{\tau_0}}$.

To deduce convergence rates for numerical schemes in later chapters on a uniform partition I_τ, we introduce a discrete version $E_{h,\tau}(\cdot) \equiv \{E_{h,\tau}(t) \mid t \geq 0\}$ of the semigroup $E_h(\cdot)$ via

$$E_{h,\tau}(t) = A_0^n \qquad \forall\, t \in [t_{n-1}, t_n) \qquad n = 1, 2, \cdots, N\,,$$

where $A_0 = (\mathbb{1}_h - \tau\Delta_h)^{-1}$, and set

$$G_\tau(\cdot) \triangleq E_h(\cdot)\Pi_h - E_{h,\tau}(\cdot)\Pi_h\,.$$

The following two lemmata bound the distance between $E_h(\cdot)$ and $E_{h,\tau}(\cdot)$.

Lemma 2.4 *There exists a constant C, such that for any $t \in (0,T]$, $h\,, \tau \in (0,1]$,*

$$\begin{cases} \|G_\tau(t)x\| \leq C\tau \|\Pi_h x\|_{\dot{\mathbb{H}}_h^2} & \forall\, x \in \dot{\mathbb{H}}^2\,, \qquad (2.10a)\\ \|G_\tau(t)x\| \leq C\tau^{\gamma/2} t^{-(\gamma-1)/2} \|x\|_{\dot{\mathbb{H}}^1} & \forall\, x \in \dot{\mathbb{H}}^1\,, \gamma \in [1,2]\,, \qquad (2.10b)\\ \|G_\tau(t)x\| \leq C\tau^{\gamma/2} t^{-\gamma/2} \|x\| & \forall\, x \in \mathbb{L}^2\,, \gamma \in [0,2]\,. \qquad (2.10c) \end{cases}$$

Proof For any $t \in [t_{n-1}, t_n)$, consider the identity

$$G_\tau(t)x = \big[E_h(t)\Pi_h - E_h(t_n)\Pi_h\big]x + \big[E_h(t_n)\Pi_h - A_0^n\Pi_h\big]x\,. \tag{2.11}$$

By applying Remark 2.1 and [7, Lemma 3.3 (iv)], we have

$$\begin{aligned} \|G_\tau(t)x\| &\leq \big\|\big[E_h(t)\Pi_h - E_h(t_n)\Pi_h\big]x\big\| + \big\|\big[E_h(t_n)\Pi_h - A_0^n\Pi_h\big]x\big\| \\ &\leq C\tau\|\Pi_h x\|_{\dot{\mathbb{H}}_h^2}\,, \end{aligned}$$

which is the assertion (2.10a). By Remark 2.1 and [8, Lemma 1], it follows for the first term of (2.11) that

$$\begin{aligned} &\big\|\big[E_h(t)\Pi_h - E_h(t_n)\Pi_h\big]x\big\| \\ &\quad = \big\|(-\Delta_h)^{-\gamma/2}\big[\mathbb{1}_h - E_h(t_n - t)\big](-\Delta_h)^{(\gamma-1)/2}E_h(t)(-\Delta_h)^{1/2}\Pi_h x\big\| \\ &\quad \leq C(t_n - t)^{\gamma/2} t^{-(\gamma-1)/2}\|x\|_{\dot{\mathbb{H}}_h^1} \end{aligned}$$

$$\leq C\tau^{\gamma/2} t^{-(\gamma-1)/2} \|x\|_{\dot{\mathbb{H}}^1} \,.$$

Relying on [7, Lemma 3.3 (iii), (iv)] and applying an interpolation argument, we have

$$\big\| \big[E_h(t_n)\Pi_h - A_0^n \Pi_h\big]x \big\| \leq C\tau^{\gamma/2} t_n^{-(\gamma-1)/2} \|x\|_{\dot{\mathbb{H}}^1} \leq C\tau^{\gamma/2} t^{-(\gamma-1)/2} \|x\|_{\dot{\mathbb{H}}^1} \,.$$

A combination of these two estimates and (2.11) then leads to the assertion (2.10b).

The third assertion (2.10c) can be derived in the same way. □

2.2 Bounds in Strong Norms for Linear SPDEs and BSPDEs

A relevant part of the analysis of Problem **(SLQ)** is to study related linear SPDEs and BSPDEs on a given filtered probability space $(\Omega, \mathcal{F}, \mathbb{F}, \mathbb{P})$. In the below, we present definitions of their strong and mild solutions, and show their stability properties in strong norms.

a. Linear SPDEs. We begin with the following SPDE ($\beta \in \mathbb{R}$)

$$\begin{cases} \mathrm{d}X(t) = \big[\Delta X(t) + f(t)\big]\,\mathrm{d}t + \big[\beta X(t) + g(t)\big]\,\mathrm{d}W(t) \qquad t \in (0, T]\,, \\ X(0) = x\,, \end{cases} \tag{2.12}$$

where $f(\cdot)\,, g(\cdot)$ are $\mathbb{F}$-adapted processes, and $x \in \mathbb{L}^2$.

Definition 2.1 **(a)** An $\mathbb{F}$-adapted, continuous stochastic process $X(\cdot)$ is called a *strong solution* to (2.12) if

(1) $X(\cdot) \in \mathbb{H}_0^1 \cap \mathbb{H}^2$ for a.e. $(t\,, \omega) \in [0, T] \times \Omega$ and $\Delta X(\cdot) \in L^1(0, T; \mathbb{L}^2)$ a.s.;
(2) For all $t \in [0, T]$,

$$X(t) = x + \int_0^t \big[\Delta X(s) + f(s)\big]\,\mathrm{d}s + \int_0^t \big[\beta X(s) + g(s)\big]\,\mathrm{d}W(s) \qquad \text{a.s.}$$

(b) An $\mathbb{F}$-adapted, continuous stochastic process $X(\cdot)$ is called a *mild solution* to (2.12) if $X(\cdot) \in L^2_{\mathbb{F}}(0, T; \mathbb{L}^2)$, and for all $t \in [0, T]$,

$$X(t) = E(t)x + \int_0^t E(t-s) f(s)\,\mathrm{d}s + \int_0^t E(t-s)\big[\beta X(s) + g(s)\big]\,\mathrm{d}W(s) \qquad \text{a.s.}$$

If $X(\cdot)$ is a strong solution to (2.12), then $X(\cdot)$ is also a mild solution to (2.12); see e.g. [9, 10]. The following result is on the regularity of (2.12).

Lemma 2.5 *Let* $x \in \mathbb{H}_0^1 \cap \mathbb{H}^2$, $f(\cdot) \in L^2_{\mathbb{F}}(0, T; \mathbb{H}_0^1)$ *and* $g(\cdot) \in L^2_{\mathbb{F}}(0, T; \mathbb{H}_0^1 \cap \mathbb{H}^2)$. *Then* (2.12) *admits a unique strong solution, and there exists a constant C such that*

$$\begin{aligned}&\mathbb{E}\Big[\sup_{t\in[0,T]}\|X(t)\|_{\dot{\mathbb{H}}^\gamma}^2\Big]+\mathbb{E}\Big[\int_0^T\|X(t)\|_{\dot{\mathbb{H}}^{\gamma+1}}^2\,\mathrm{d}t\Big]\\&\leq C\Big\{\|x\|_{\dot{\mathbb{H}}^\gamma}^2+\mathbb{E}\Big[\int_0^T\|f(t)\|_{\dot{\mathbb{H}}^{\gamma-1}}^2+\|g(t)\|_{\dot{\mathbb{H}}^\gamma}^2\,\mathrm{d}t\Big]\Big\}\qquad\gamma=-1,0,1,2\,.\end{aligned}\tag{2.13}$$

Furthermore, if $g(\cdot)\in C_{\mathbb{F}}\big([0,T];L^2(\Omega;\mathbb{L}^2)\big)$, *then for any* t , $s\in[0,T]$ *it holds that*

$$\begin{aligned}&\mathbb{E}\big[\|X(t)-X(s)\|^2\big]\\&\leq C|t-s|\big[\|x\|_{\mathbb{H}_0^1}^2+\|f(\cdot)\|_{L^2_{\mathbb{F}}(0,T;\mathbb{H}_0^1)}^2+\|g(\cdot)\|_{L^2_{\mathbb{F}}(0,T;\mathbb{H}_0^1)\cap C_{\mathbb{F}}([0,T];L^2(\Omega;\mathbb{L}^2))}^2\big]\,.\end{aligned}\tag{2.14}$$

Proof (1) Verification of (2.13). Firstly, [10, Theorem 3.14] implies that (2.12) admits a unique mild solution. Then we adopt the spectral decomposition technique to derive the existence of a (unique) strong solution and validate assertion (2.13); see e.g. [11, Chap. 6]. Specifically, by applying eigenvalues $\{\lambda_n\}_{n\geq1}$ and eigenfunctions $\{\varphi_n\}_{n\geq1}$ of Δ, and setting

$$x=\sum_{i=1}^\infty x_{i,0}\varphi_i\,,\qquad f(\cdot)=\sum_{i=1}^\infty f_i(\cdot)\varphi_i\,,\qquad g(\cdot)=\sum_{i=1}^\infty g_i(\cdot)\varphi_i\,,$$

we have a family of $\mathbb{R}$-valued stochastic differential equations (SDEs, for short) instead of (2.12)

$$\begin{cases}\mathrm{d}x_i(t)=\big[-\lambda_i x_i(t)+f_i(t)\big]\,\mathrm{d}t+\big[\beta x_i(t)+g_i(t)\big]\,\mathrm{d}W(t)\qquad t\in(0,T]\,,\\ x_i(0)=x_{i,0}\,.\end{cases}\tag{2.15}$$

By standard SDE theory, system (2.15) admits a unique solution, with each $x_i(\cdot)\in L^2_{\mathbb{F}}\big(\Omega;C([0,T];\mathbb{R})\big)$, and

$$\begin{aligned}&\mathbb{E}\Big[\sup_{t\in[0,T]}|x_i(t)|^2\Big]+\mathbb{E}\Big[\int_0^T\lambda_i|x_i(t)|^2\,\mathrm{d}t\Big]\\&\leq C\Big\{|x_{i,0}|^2+\mathbb{E}\Big[\int_0^T\frac{1}{\lambda_i}|f_i(t)|^2+|g_i(t)|^2\,\mathrm{d}t\Big]\Big\}\,,\end{aligned}$$

where the constant C is independent of λ_i .

By setting

$$X(\cdot)=\sum_{i=1}^\infty x_i(\cdot)\varphi_i\,,$$

we find that $X(\cdot)$ is the mild solution to (2.12) and by Fatou's Lemma for $\gamma=-1,0,1,2$

$$\mathbb{E}\Big[\sup_{t\in[0,T]}\|X(t)\|_{\dot{\mathbb{H}}^\gamma}^2\Big]+\mathbb{E}\Big[\int_0^T\|X(t)\|_{\dot{\mathbb{H}}^{\gamma+1}}^2\,\mathrm{d}t\Big]$$

$$\leq \sum_{i=1}^{\infty} \lambda_i^\gamma \mathbb{E}\Big[\sup_{t\in[0,T]} |x_i(t)|^2\Big] + \sum_{i=1}^{\infty} \lambda_i^\gamma \mathbb{E}\Big[\int_0^T \lambda_i |x_i(t)|^2 \,\mathrm{d}t\Big]$$

$$\leq C \sum_{i=1}^{\infty} \Big\{\lambda_i^\gamma |x_{i,0}|^2 + \mathbb{E}\Big[\int_0^T \lambda_i^{\gamma-1} |f_i(t)|^2 + \lambda_i^\gamma |g_i(t)|^2 \,\mathrm{d}t\Big]\Big\}$$

$$\leq C\Big\{\|x\|^2_{\dot{\mathbb{H}}^\gamma} + \mathbb{E}\Big[\int_0^T \|f(t)\|^2_{\dot{\mathbb{H}}^{\gamma-1}} + \|g(t)\|^2_{\dot{\mathbb{H}}^\gamma} \,\mathrm{d}t\Big]\Big\},$$

from which we may conclude that $X(\cdot)$ is in fact a strong solution. Moreover, it settles the assertion (2.13) then.

(2) Verification of (2.14). Starting with (2.12), for any $t, s \in [0, T]$ with $s \leq t$, as well as Hölder's inequality and the Itô isometry (see e.g. [12]), we have

$$\begin{aligned}
&\mathbb{E}\big[\|X(t) - X(s)\|^2\big] \\
&\leq C\Big\{\big\|\big[E(t) - E(s)\big]x\big\|^2 + \mathbb{E}\Big[\int_0^s \big\|E(s-\theta)\big[E(t-s) - \mathbb{1}\big]f(\theta)\big\|^2 \,\mathrm{d}\theta\Big] \\
&\quad + |t-s|\mathbb{E}\Big[\int_s^t \big\|E(t-\theta)f(\theta)\big\|^2 \,\mathrm{d}\theta\Big] \\
&\quad + \mathbb{E}\Big[\int_0^s \big\|E(s-\theta)\big[E(t-s) - \mathbb{1}\big]\big[\beta X(\theta) + g(\theta)\big]\big\|^2 \,\mathrm{d}\theta\Big] \\
&\quad + \mathbb{E}\Big[\int_s^t \big\|E(t-\theta)\big[\beta X(\theta) + g(\theta)\big]\big\|^2 \,\mathrm{d}\theta\Big]\Big\} \\
&=: C \sum_{i=1}^{5} I_i\,.
\end{aligned}$$

By (2.2b) in Lemma 2.1, it follows that

$$I_1 \leq \big\|E(s)(-\Delta)^{-1/2}\big[E(t-s) - \mathbb{1}\big](-\Delta)^{1/2}x\big\|^2 \leq C|t-s|\|x\|^2_{\mathbb{H}_0^1}\,.$$

By (2.13) with $\gamma = 0$, and the assumption on $g(\cdot)$, we have

$$I_3 + I_5 \leq C|t-s|\big[\|x\|^2 + \|f(\cdot)\|^2_{L^2_{\mathbb{F}}(0,T;\mathbb{L}^2)} + \|g(\cdot)\|^2_{C_{\mathbb{F}}([0,T];L^2(\Omega;\mathbb{L}^2))}\big]\,.$$

Similar to the estimation of I_1, we may use Lemma 2.1 and (2.13) with $\gamma = 0$ to conclude

$$\begin{aligned}
I_2 + I_4 &\leq C|t-s|\big[\|f(\cdot)\|^2_{L^2_{\mathbb{F}}(0,T;\mathbb{H}_0^1)} + \|X(\cdot)\|^2_{L^2_{\mathbb{F}}(0,T;\mathbb{H}_0^1)} + \|g(\cdot)\|^2_{L^2_{\mathbb{F}}(0,T;\mathbb{H}_0^1)}\big] \\
&\leq C|t-s|\big[\|x\|^2 + \|f(\cdot)\|^2_{L^2_{\mathbb{F}}(0,T;\mathbb{H}_0^1)} + \|g(\cdot)\|^2_{L^2_{\mathbb{F}}(0,T;\mathbb{H}_0^1)}\big]\,.
\end{aligned}$$

That completes the proof of (2.14). □

b. Linear BSPDEs. To handle the adjoint equation coming from Pontryagin's maximum principle which is related to Problem **(SLQ)**, we consider the following linear BSPDE ($\beta \in \mathbb{R}$)

$$\begin{cases} \mathrm{d}Y(t) = \big[-\Delta Y(t) - \beta Z(t) + f(t) \big]\, \mathrm{d}t + Z(t)\, \mathrm{d}W(t) & \forall t \in [0, T)\,, \\ Y(T) = Y_T\,. \end{cases} \tag{2.16}$$

Definition 2.2 **(a)** An $\mathbb{L}^2 \times \mathbb{L}^2$-valued stochastic process tuple $\big(Y(\cdot), Z(\cdot)\big)$ is called a *strong solution* to (2.16) if

(1) $Y(\cdot)$ is $\mathbb{F}$-adapted and continuous, and $Z(\cdot) \in L^2_{\mathbb{F}}(0, T; \mathbb{L}^2)$;
(2) $Y(\cdot) \in \mathbb{H}^1_0 \cap \mathbb{H}^2$ for a.e.$(t\,, \omega) \in [0, T] \times \Omega$ and $\Delta Y(\cdot) \in L^1(0, T; \mathbb{L}^2)$ a.s.;
(3) For all $t \in [0, T]$,

$$Y(t) = Y_T + \int_t^T \big[\Delta Y(s) + \beta Z(s) - f(s)\big]\, \mathrm{d}s - \int_t^T Z(s)\, \mathrm{d}W(s) \qquad \text{a.s.}$$

(b) An $\mathbb{L}^2 \times \mathbb{L}^2$-valued stochastic process tuple $\big(Y(\cdot), Z(\cdot)\big)$ is called a *mild solution* to (2.16) if

(1) $Y(\cdot)$ is $\mathbb{F}$-adapted and continuous, and $Z(\cdot) \in L^2_{\mathbb{F}}(0, T; \mathbb{L}^2)$;
(2) For all $t \in [0, T]$,

$$\begin{aligned} Y(t) = E(T-t)Y_T &+ \int_t^T E(s-t)\big[\beta Z(s) - f(s)\big]\, \mathrm{d}s \\ &- \int_t^T E(s-t) Z(s)\, \mathrm{d}W(s) \qquad \text{a.s.} \end{aligned}$$

Similar to SPDEs, if $\big(Y(\cdot), Z(\cdot)\big)$ is a strong solution to (2.16), then $\big(Y(\cdot), Z(\cdot)\big)$ is also a mild solution to (2.16); see e.g. [10, 13].

The following lemma can be derived by a spectral decomposition argument as in the proof of Lemma 2.5.

Lemma 2.6 *Assume $Y_T \in \mathbb{H}^1_0 \cap \mathbb{H}^2$, and $f(\cdot) \in L^2_{\mathbb{F}}(0, T; \mathbb{H}^1_0)$. Then* (2.16) *admits a unique strong solution $\big(Y(\cdot), Z(\cdot)\big) \in \big(L^2_{\mathbb{F}}(\Omega; C([0, T]; \mathbb{H}^1_0)) \cap L^2_{\mathbb{F}}(0, T; \mathbb{H}^1_0 \cap \mathbb{H}^2)\big) \times L^2_{\mathbb{F}}(0, T; \mathbb{H}^1_0)$. Furthermore, there exists a constant C such that*

$$\begin{aligned} &\mathbb{E}\Big[\sup_{t\in[0,T]} \|Y(t)\|^2_{\dot{\mathbb{H}}^\gamma}\Big] + \mathbb{E}\Big[\int_0^T \|Y(t)\|^2_{\dot{\mathbb{H}}^{\gamma+1}} + \|Z(t)\|^2_{\dot{\mathbb{H}}^\gamma}\, \mathrm{d}t\Big] \\ &\qquad \le C\,\mathbb{E}\Big[\|Y_T\|^2_{\dot{\mathbb{H}}^\gamma} + \int_0^T \|f(t)\|^2_{\dot{\mathbb{H}}^{\gamma-1}}\, \mathrm{d}t\Big] \qquad \gamma = -1, 0, 1, 2\,. \end{aligned} \tag{2.17}$$

2.3 Open-Loop and Closed-Loop Control Strategies for SLQ Problems

a. Pontryagin's maximum principle. At the center of the 'open-loop approach' stands *Pontryagin's maximum principle* (see e.g. [10]), which involves the further adjoint processes $\big(Y(\cdot), Z(\cdot)\big) \in \big(L^2_{\mathbb{F}}(\Omega; C([0,T]; \mathbb{H}^1_0)) \cap L^2_{\mathbb{F}}(0,T; \mathbb{H}^1_0 \cap \mathbb{H}^2)\big) \times L^2_{\mathbb{F}}(0,T; \mathbb{H}^1_0)$ next to the optimal pair $\big(X^*(\cdot), U^*(\cdot)\big)$. While $Y(\cdot)$ is an adjoint process as in the context of necessary optimality conditions for deterministic problems, the additional $Z(\cdot)$ makes sure that it is $\mathbb{F}$-adapted as well. Consequently, we consider the quadruple $\big(X^*(\cdot), Y(\cdot), Z(\cdot), U^*(\cdot)\big)$ that solves

$$\begin{cases} \mathrm{d}X^*(t) = \big[\Delta X^*(t) + U^*(t)\big]\,\mathrm{d}t + \big[\beta X^*(t) + \sigma(t)\big]\mathrm{d}W(t) & t \in (0,T]\,, \quad (2.18a) \\ \mathrm{d}Y(t) = \big[-\Delta Y(t) - \beta Z(t) + X^*(t)\big]\,\mathrm{d}t + Z(t)\,\mathrm{d}W(t) & t \in [0,T)\,, \quad (2.18b) \\ X^*(0) = x\,, \quad Y(T) = -\alpha X^*(T)\,, & \quad (2.18c) \end{cases}$$

together with the optimality condition

$$U^*(t) - Y(t) = 0 \qquad t \in [0,T]\,. \tag{2.19}$$

b. Stochastic Riccati equation and feedback control. By LQ theory, Problem **(SLQ)** admits a unique optimal pair $\big(X^*(\cdot), U^*(\cdot)\big)$, which is linked by an operator-valued function $\mathcal{P}(\cdot)$, which solves the *stochastic Riccati equation*

$$\begin{cases} \mathcal{P}'(t) + \Delta\mathcal{P}(t) + \mathcal{P}(t)\Delta + \beta^2\mathcal{P}(t) + \mathbb{1} - \mathcal{P}^2(t) = 0 & t \in [0,T]\,, \\ \mathcal{P}(T) = \alpha\mathbb{1}\,. \end{cases} \tag{2.20}$$

The following definition clarifies the notion of a solution to this deterministic terminal value problem.

Definition 2.3 We call $\mathcal{P}(\cdot) \in C\big([0,T]; \mathbb{S}(\mathbb{L}^2)\big)$ a *mild solution* to (2.20) if for any $x \in \mathbb{L}^2$ and any $s \in [0,T]$,

$$\mathcal{P}(s)x = \alpha E\big(2(T-s)\big)x + \int_s^T E(\theta - s)\big[\beta^2\mathcal{P}(\theta) + \mathbb{1} - \mathcal{P}^2(\theta)\big]E(\theta - s)x\,\mathrm{d}\theta\,. \tag{2.21}$$

It is shown in [14, Theorem 2.2] that (2.20) admits a unique mild solution $\mathcal{P}(\cdot) \in C\big([0,T]; \mathbb{S}(\mathbb{L}^2)\big)$; furthermore, by

$$\begin{aligned} &\inf_{U(\cdot) \in L^2_{\mathbb{F}}(t,T;\mathbb{L}^2)} \mathbb{E}\Big[\int_t^T \big[\|X\big(s; t, x, U(\cdot)\big)\|^2 + \|U(s)\|^2\big]\,\mathrm{d}t + \alpha\|X\big(T; t, x, U(\cdot)\big)\|^2\Big] \\ &\quad = (\mathcal{P}(t)x, x)_{\mathbb{L}^2}\,, \end{aligned}$$

we actually deduce that $\mathcal{P}(\cdot) \in C\big([0,T];\mathbb{S}_+(\mathbb{L}^2)\big)$, where $X\big(\cdot;t,x,U(\cdot)\big)$ solves

$$\begin{cases} \mathrm{d}X(s) = \big[\Delta X(s) + U(s)\big]\mathrm{d}t + \beta X(s)\,\mathrm{d}W(s) & s \in (t,T]\,, \\ X(t) = x\,. \end{cases}$$

Relying on $\mathcal{P}(\cdot)$ and $\eta(\cdot)$, where the latter solves the $\mathcal{P}(\cdot)$-dependent deterministic terminal value PDE (1.10), we may then characterize the solution $U^*(\cdot)$ of Problem **(SLQ)** by the following *feedback control law*

$$U^*(t) = -\mathcal{P}(t)X^*(t) - \eta(t) \qquad t \in [0,T]\,. \tag{2.22}$$

This feedback law (2.22), together with the optimality condition (2.19) and the results in Sect. 2.2, allows us to verify improved regularity properties and bounds for the optimal pair $\big(X^*(\cdot), U^*(\cdot)\big)$; see Lemma 2.7 in this section. Besides, (2.22) is crucial to algorithmically settle the 'closed-loop approach', which is proposed in Chap. 4.

With the help of the preceding two parts, we can now show bounds in stronger norms for the solution of FBSPDE (2.18)–(2.19)—and hence of Problem **(SLQ)**.

Lemma 2.7 *Suppose that* **(A)** *holds. Then* $\big(X^*(\cdot), U^*(\cdot)\big)$*, the optimal pair of Problem* **(SLQ)***, belongs to* $L^2_{\mathbb{F}}\big(\Omega; C([0,T];\mathbb{H}^1_0)\big) \times L^2_{\mathbb{F}}\big(\Omega; C([0,T];\mathbb{H}^1_0)\big)$*, and there exists a constant* C *such that*

$$\begin{cases} \sup\limits_{t\in[0,T]} \Big(\mathbb{E}\big[\|U^*(t)\|^4\big]\Big)^{1/2} \le C\big[\|x\|^2 + \|\sigma(\cdot)\|^2_{C([0,T];\mathbb{L}^2)}\big]\,, & (2.23a) \\ \mathbb{E}\Big[\sup\limits_{t\in[0,T]} \|U^*(t)\|^2 + \displaystyle\int_0^T \|U^*(t)\|^2_{\mathbb{H}^1_0}\,\mathrm{d}t\Big] \le C\big[\|x\|^2 + \|\sigma(\cdot)\|^2_{L^2(0,T;\mathbb{L}^2)}\big]\,, & (2.23b) \\ \mathbb{E}\Big[\sup\limits_{t\in[0,T]} \|X^*(t)\|^2_{\mathbb{H}^1_0} + \displaystyle\int_0^T \|X^*(t)\|^2_{\mathbb{H}^1_0\cap\mathbb{H}^2}\,\mathrm{d}t\Big] & \\ \qquad\qquad \le C\big[\|x\|^2_{\mathbb{H}^1_0} + \|\sigma(\cdot)\|^2_{L^2(0,T;\mathbb{H}^1_0)}\big]\,. & (2.23c) \end{cases}$$

Furthermore, there exists a constant C *such that for any* $s, t \in [0,T]$

$$\mathbb{E}\big[\|U^*(t) - U^*(s)\|^2\big] \le C|t-s|\big[\|x\|^2_{\mathbb{H}^1_0} + \|\sigma\|^2_{C([0,T];\mathbb{H}^1_0)}\big]\,. \tag{2.24}$$

Proof (1) Verification of (2.23a). The fact that $\mathcal{P}(\cdot) \in C\big([0,T];\mathbb{S}(\mathbb{L}^2)\big)$ leads to $\sup_{t\in[0,T]} \|P(t)\|_{\mathcal{L}(\mathbb{L}^2)} \le C$. Then by equation (1.10), we may use Gronwall's inequality to derive

$$\sup_{t\in[0,T]} \|\eta(t)\|^2 + \int_0^T \big[\|\eta(t)\|^2_{\mathbb{H}^1_0} + \big(\mathcal{P}(t)\eta(t), \eta(t)\big)_{\mathbb{L}^2}\big]\mathrm{d}t \le C\int_0^T \|\sigma(t)\|^2\,\mathrm{d}t\,. \tag{2.25}$$

Here we apply the fact that $\mathcal{P}(\cdot) \in C\big([0,T];\mathbb{S}_+(\mathbb{L}^2)\big)$. Then, by inserting the feedback law (2.22) into SPDE (2.18a) and applying Gronwall's inequality as well as the

Burkholder-Davis-Gundy inequality (BDG inequality, for short), we can see that

$$\begin{aligned}&\mathbb{E}\Big[\sup_{t\in[0,T]}\|X^*(t)\|^2\Big]+\mathbb{E}\Big[\int_0^T\|X^*(t)\|_{\mathbb{H}_0^1}^2+\big(\mathcal{P}(t)X^*(t),X^*(t)\big)_{\mathbb{L}^2}\,\mathrm{d}t\Big]\\&\leq C\Big[\|x\|^2+\int_0^T\|\eta(t)\|^2+\|\sigma(t)\|^2\,\mathrm{d}t\Big]\\&\leq C\big[\|x\|^2+\|\sigma(\cdot)\|_{L^2(0,T;\mathbb{L}^2)}^2\big]. \end{aligned}\tag{2.26}$$

On the other hand, we may apply Itô's formula to $\|X^*(t)\|^4$, the feedback law (2.22), and the estimate (2.25) for $\eta(\cdot)$ to get

$$\begin{aligned}\sup_{t\in[0,T]}\mathbb{E}\big[\|X^*(t)\|^4\big]&\leq C\Big[\|x\|^4+\int_0^T\|\eta(t)\|^4+\|\sigma(t)\|^4\,\mathrm{d}t\Big]\\&\leq C\big[\|x\|^2+\|\sigma(\cdot)\|_{C([0,T];\mathbb{L}^2)}^2\big]^2.\end{aligned}$$

Using (2.22) once more, we can derive the assertion (2.23a).

(2) Verification of (2.23b). On one hand, we use the feedback law (2.22) and (2.25), (2.26) to get

$$\begin{aligned}\mathbb{E}\Big[\sup_{t\in[0,T]}\|U^*(t)\|^2\Big]&\leq C\Big\{\sup_{t\in[0,T]}\|\mathcal{P}(t)\|_{\mathcal{L}(\mathbb{L}^2)}^2\mathbb{E}\Big[\sup_{t\in[0,T]}\|X^*(t)\|^2\Big]+\sup_{t\in[0,T]}\|\eta(t)\|^2\Big\}\\&\leq C\big[\|x\|^2+\|\sigma(\cdot)\|_{L^2(0,T;\mathbb{L}^2)}^2\big].\end{aligned}$$

On the other hand, the optimality condition (2.19), Lemma 2.6 and (2.26), yields

$$\begin{aligned}\mathbb{E}\Big[\int_0^T\|U^*(t)\|_{\mathbb{H}_0^1}^2\,\mathrm{d}t\Big]&=\mathbb{E}\Big[\int_0^T\|Y(t)\|_{\mathbb{H}_0^1}^2\,\mathrm{d}t\Big]\\&\leq C\,\mathbb{E}\Big[\|X^*(T)\|^2\Big]+\int_0^T\|X^*(t)\|^2\,\mathrm{d}t\Big]\\&\leq C\big[\|x\|^2+\|\sigma(\cdot)\|_{L^2(0,T;\mathbb{L}^2)}^2\big].\end{aligned}$$

That settles assertion (2.23b).

(3) Verification of (2.23c). By Lemma 2.5 with $f(\cdot)=U^*(\cdot)$, $g(\cdot)=\sigma(\cdot)$ and (2.23b), we have

$$\begin{aligned}&\mathbb{E}\Big[\sup_{t\in[0,T]}\|X^*(t)\|_{\mathbb{H}_0^1}^2\Big]+\mathbb{E}\Big[\int_0^T\|X^*(t)\|_{\mathbb{H}_0^1\cap\mathbb{H}^2}^2\,\mathrm{d}t\Big]\\&\leq C\Big\{\|x\|_{\mathbb{H}_0^1}^2+\mathbb{E}\Big[\int_0^T\|U^*(t)\|^2+\|\sigma(t)\|_{\mathbb{H}_0^1}^2\,\mathrm{d}t\Big]\Big\}\\&\leq C\big[\|x\|_{\mathbb{H}_0^1}^2+\|\sigma(\cdot)\|_{L^2(0,T;\mathbb{H}_0^1)}^2\big].\end{aligned}$$

(4) Verification of (2.24). For any $t \in [0, T]$, we find that $X^*(t) \in \mathbb{D}^{1,2}(\mathbb{L}^2)$ (see, e.g. [4]), and $D_\theta X^*(\cdot)$ solves

$$\begin{cases} \mathrm{d}D_\theta X^*(t) = \big[\Delta - \mathcal{P}(t)\big] D_\theta X^*(t)\,\mathrm{d}t + \beta D_\theta X^*(t)\,\mathrm{d}W(t) \qquad t \in (\theta, T]\,, \\ D_\theta X^*(\theta) = \beta X^*(\theta) + \sigma(\theta)\,. \end{cases} \tag{2.27}$$

By [15, Proposition 3.2], the Malliavin derivative $\big(D_\theta Y(\cdot), D_\theta Z(\cdot)\big)$ of the solution $\big(Y(\cdot), Z(\cdot)\big)$ to the BPSDE (2.18b) satisfies

$$\begin{cases} \mathrm{d}D_\theta Y(t) = \big[-\Delta D_\theta Y(t) - \beta D_\theta Z(t) + D_\theta X^*(t)\big]\mathrm{d}t \\ \qquad\qquad +D_\theta Z(t)\mathrm{d}W(t) \qquad t \in [\theta, T)\,, \\ D_\theta Y(T) = -\alpha D_\theta X^*(T)\,, \\ D_\theta Y(t) = 0\,, D_\theta Z(t) = 0 \qquad t \in [0, \theta)\,, \end{cases} \tag{2.28}$$

and

$$Z(t) = D_t Y(t) \qquad \text{a.e.}\, t \in [0, T]\,. \tag{2.29}$$

Now Lemmata 2.6 and 2.5 imply that

$$\begin{aligned} &\sup_{\theta\in[0,T]} \bigg\{ \sup_{t\in[\theta,T]} \mathbb{E}\big[\|D_\theta Y(t)\|^2\big] + \mathbb{E}\Big[\int_\theta^T \|\nabla D_\theta Y(t)\|^2 + \|D_\theta Z(t)\|^2\,\mathrm{d}t\Big]\bigg\} \\ &\quad\le C \sup_{\theta\in[0,T]} \mathbb{E}\Big[\alpha^2\|D_\theta X^*(T)\|^2 + \int_\theta^T \|D_\theta X^*(t)\|^2\,\mathrm{d}t\Big] \\ &\quad\le C \sup_{\theta\in[0,T]} \mathbb{E}\big[\|X^*(\theta)\|^2 + \|\sigma(\theta)\|^2\big] \\ &\quad\le C\big[\|x\|^2 + \|\sigma(\cdot)\|^2_{C([0,T];\mathbb{L}^2)}\big]\,. \end{aligned} \tag{2.30}$$

By the optimality condition (2.19), (2.17), (2.29) and (2.30), we conclude that for $s \le t$,

$$\begin{aligned} &\mathbb{E}\big[\|U^*(t) - U^*(s)\|^2\big] \\ &\quad\le C(t-s)\mathbb{E}\Big[\int_0^T \|Y(t)\|^2_{\mathbb{H}_0^1\cap\mathbb{H}^2} + \|Z(t)\|^2 + \|X^*(t)\|^2\,\mathrm{d}t\Big] \\ &\qquad + C\mathbb{E}\Big[\int_s^t \|D_\theta Y(\theta)\|^2\,\mathrm{d}\theta\Big] \\ &\quad\le C|t-s|\bigg\{\mathbb{E}\Big[\|X^*(T)\|^2_{\mathbb{H}_0^1} + \int_0^T \|X^*(t)\|^2\,\mathrm{d}t\Big] + \sup_{\theta\in[0,T]} \mathbb{E}\big[\|D_\theta Y(\theta)\|^2\big]\bigg\} \\ &\quad\le C|t-s|\big[\|x\|^2_{\mathbb{H}_0^1} + \|\sigma(\cdot)\|^2_{C([0,T];\mathbb{L}^2)\cap L^2(0,T;\mathbb{H}_0^1)}\big]\,. \end{aligned}$$

That settles assertion (2.24). □

References

1. H. Fujita, T. Suzuki, Evolution problems, in *Finite element methods. Part 1*, ed. by P.G. Ciarlet, J.-L. Lions, vol. II of Handbook of Numerical Analysis, North-Holland, Amsterdam, pp. 789–928 (1991)
2. R. Kruse, Optimal error estimates of Galerkin finite element methods for stochastic partial differential equations with multiplicative noise. IMA J. Numer. Anal. **34**, 217–251 (2014)
3. A. Pazy, Semigroups of linear operators and applications to partial differential equations. *Applied Mathematical Sciences*, vol. 44 (Springer, New York, 1983)
4. D. Nualart, The Malliavin calculus and related topics. *Probability and its Applications (New York)*, 2nd ed. (Springer, Berlin, 2006)
5. S.C. Brenner, L.R. Scott, The mathematical theory of finite element methods, *Texts in Applied Mathematics*, vol. 15, 3rd ed. (Springer, New York, 2008)
6. V. Thomée, Galerkin finite element methods for parabolic problems. *Springer Series in Computational Mathematics*, vol. 25, 2nd edn. (Springer, Berlin, 2006)
7. A. Tambue, J.D. Mukam, Strong convergence of the linear implicit Euler method for the finite element discretization of semilinear SPDEs driven by multiplicative or additive noise. Appl. Math. Comput. **346**, 23–40 (2019)
8. J.D. Mukam, A. Tambue, Strong convergence analysis of the stochastic exponential Rosenbrock scheme for the finite element discretization of semilinear SPDEs driven by multiplicative and additive noise. J. Sci. Comput. **74**, 937–978 (2018)
9. C. Prévôt, M. Röckner, A concise course on stochastic partial differential equations. *Lecture Notes in Mathematics*, vol. 1905 (Springer, Berlin, 2007)
10. Q. Lü, X. Zhang, Mathematical control theory for stochastic partial differential equations. *Probability Theory and Stochastic Modelling*, vol. 101 (Springer, Cham, 2021)
11. P.-L. Chow, Stochastic partial differential equations. *Advances in Applied Mathematics*, 2nd ed. (CRC Press, Boca Raton, FL, 2015)
12. G. Da Prato, J. Zabczyk, Stochastic equations in infinite dimensions. *Encyclopedia of Mathematics and its Applications*, vol. 44 (Cambridge University Press, Cambridge, 1992)
13. A.R. Al-Hussein, Strong, mild and weak solutions of backward stochastic evolution equations. Random Oper. Stochastic Equs. **13**, 129–138 (2005)
14. Q. Lü, Well-posedness of stochastic Riccati equations and closed-loop solvability for stochastic linear quadratic optimal control problems. J. Diff. Equs. **267**, 180–227 (2019)
15. F. Dou, Q. Lü, Partial approximate controllability for linear stochastic control systems. SIAM J. Control Optim. **57**, 1209–1229 (2019)

Chapter 3
Discretization Based on the Open-Loop Approach

Different analytical tools from the 'open-loop approach' (i.e., Pontryagin's maximum principle), and the 'closed-loop approach' (i.e., stochastic Riccati equation) were introduced in Sect. 2.3 to conclude improved regularity properties of the unique optimal pair $\big(X^*(\cdot), U^*(\cdot)\big)$ to Problem **(SLQ)**; in particular, they were used to verify Lemma 2.7, which contains bounds for $\big(X^*(\cdot), U^*(\cdot)\big)$ in strong norms.

This equivalent characterization of a minimizer of Problem **(SLQ)** in discrete form via the related optimality conditions or the feedback-law formulae still holds if related discretization schemes in space and time are considered: from a practical algorithmic view, however, we will see that their usability is very different. In this chapter, we focus on the 'open-loop approach', and the construction of a related numerical method.

For the numerical approach in this chapter, numerical schemes to approximate FBSPDE (2.18) with the optimality condition (2.19) are the crucial strategy point, and we refer to Fig. 3.1 for an outline; once a spatio-temporal discrete formulation of (2.18)–(2.19) has been established (see (3.70)–(3.71)) which also serves as the optimality condition for a proper discretization of Problem **(SLQ)** (see Problem $\mathbf{(SLQ)}_{h\tau}$ related to (3.68)–(3.69)), a decoupling of the BSPDE from the SPDE (after discretization) happens through a gradient decent method to simplify computations (see Algorithm 3.9). For its simulation, however, there remains to compute conditional expectations per iteration step. For the additive noise case ($\beta = 0$), an exact formulation for conditional expectations can be derived in Theorem 3.11, which leads to an algorithm that converges with optimal rate—and is free from the calculation of conditional expectations. For the multiplicative noise case, we use (data dependent) regression for its estimation in high-dimensional spaces; see Sect. 3.4.2.

This chapter starts with two preparatory sections: in Sect. 3.1, we recall shortly the derivation of convergence rates for spatio-temporal discretization schemes of SPDE (2.12) which are well-known; Sect. 3.2 addresses corresponding goals for BSPDE

© The Author(s), under exclusive license to Springer Nature Singapore Pte Ltd. 2026

A. Prohl and Y. Wang, *Numerical Methods for Optimal Control Problems with SPDEs*, SpringerBriefs on PDEs and Data Science,
https://doi.org/10.1007/978-981-95-4469-1_3

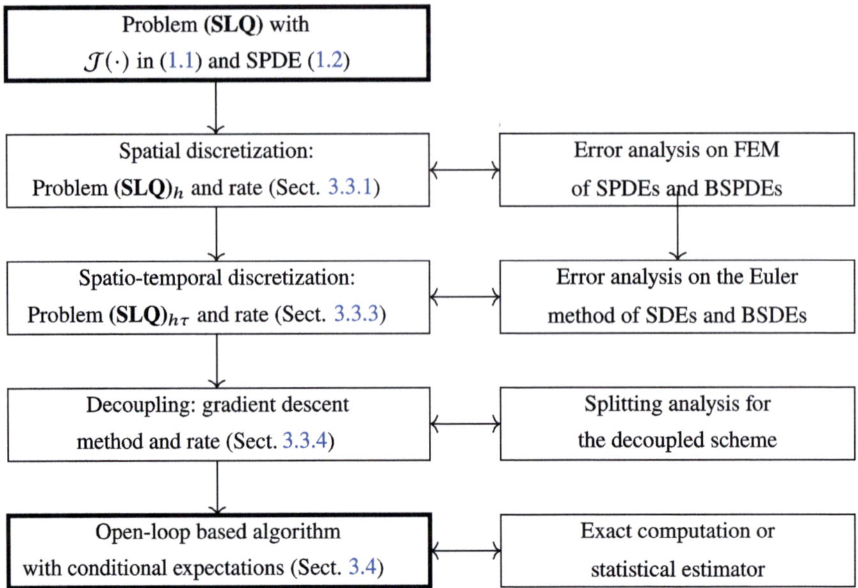

Fig. 3.1 Steps within the 'open-loop approach' to approximate Problem **(SLQ)**

(2.16), whose numerical analysis is only available in recent research articles. The results in both sections will then be used in the later ones which address Problem **(SLQ)**.

3.1 Discretization of Linear SPDEs and Rates

We split the derivation of convergence rates for the spatio-temporal discretization of SPDE (2.12) into two steps: we begin with an SDE, which results from discretization in space only; related stability bounds for the solution to SDEs may then be used in Sect. 3.1.2 to verify rates of convergence for the complete discretization in space and time.

3.1.1 Spatial Discretization and Error Estimates for Linear SPDEs

We use the notations from part **c.** of Sect. 2.1. A finite element method of (2.12) reads: Find $X_h(\cdot) \in L^2_{\mathbb{F}}\big(\Omega; C([0,T]; \mathbb{V}_h)\big)$ that satisfies

$$
\begin{cases} \mathrm{d}X_h(t) = \big[\Delta_h X_h(t) + f_h(t)\big]\,\mathrm{d}t + \big[\beta X_h(t) + g_h(t)\big]\,\mathrm{d}W(t) & t \in (0,T]\,, \\ X_h(0) = \Pi_h x\,, \end{cases} \tag{3.1}
$$

where $f_h(\cdot)\,, g_h(\cdot) \in L^2_{\mathbb{F}}(0,T;\mathbb{V}_h)$ are approximations of $f(\cdot)\,, g(\cdot)$, respectively. Note that for any $t \in [0,T]$, $X_h(t)$ takes values in the finite dimensional space $\mathbb{V}_h$. Hence problem (3.1) is a system of linear SDEs, and its solvability is obvious.

Lemma 3.1 *Suppose that* $x \in \mathbb{H}^1_0 \cap \mathbb{H}^2\,, f_h(\cdot)\,, g_h(\cdot) \in L^2_{\mathbb{F}}(0,T;\mathbb{V}_h)$*, and* $X_h(\cdot)$ *solves* (3.1)*. There exists a constant C such that*

$$
\begin{aligned}
&\mathbb{E}\Big[\sup_{t\in[0,T]} \|X_h(t)\|^2_{\dot{\mathbb{H}}^\gamma_h}\Big] + \mathbb{E}\Big[\int_0^T \|X_h(t)\|^2_{\dot{\mathbb{H}}^{\gamma+1}_h}\,\mathrm{d}t\Big] \\
&\le C\Big\{\|\Pi_h x\|^2_{\dot{\mathbb{H}}^\gamma_h} + \mathbb{E}\Big[\int_0^T \|f_h(t)\|^2_{\dot{\mathbb{H}}^{\gamma-1}_h} + \|g_h(t)\|^2_{\dot{\mathbb{H}}^\gamma_h}\,\mathrm{d}t\Big]\Big\} \quad \gamma = -1, 0, 1, 2\,.
\end{aligned} \tag{3.2}
$$

Furthermore, for a uniform partition $I_\tau = \{t_n\}_{n=0}^N \subset [0,T]$*, there exists a constant C independent of* τ *such that*

$$
\begin{aligned}
&\sum_{n=0}^{N-1} \mathbb{E}\Big[\int_{t_n}^{t_{n+1}} \|X_h(t) - X_h(t_n)\|^2_{\dot{\mathbb{H}}^\gamma_h} + \|X_h(t) - X_h(t_{n+1})\|^2_{\dot{\mathbb{H}}^\gamma_h}\,\mathrm{d}t\Big] \\
&\quad \le C\tau\Big\{\|x\|^2_{\dot{\mathbb{H}}^{\gamma+1}} + \mathbb{E}\Big[\int_0^T \|f_h(t)\|^2_{\dot{\mathbb{H}}^\gamma_h} + \|g_h(t)\|^2_{\dot{\mathbb{H}}^{\gamma+1}_h}\,\mathrm{d}t\Big]\Big\} \qquad \gamma = 0, 1\,.
\end{aligned} \tag{3.3}
$$

Proof The estimates (3.2) can be proved in the same vein as in the proof of Lemma 2.5. Here, we only bound $X_h(t) - X_h(t_n)$ on the left-hand side of (3.3); the remaining part can be derived by a similar argument. Based on the Itô isometry and (3.2), we can obtain

$$
\begin{aligned}
&\sum_{n=0}^{N-1} \mathbb{E}\Big[\int_{t_n}^{t_{n+1}} \|X_h(t) - X_h(t_n)\|^2_{\dot{\mathbb{H}}^\gamma_h}\,\mathrm{d}t\Big] \\
&\quad \le C\sum_{n=0}^{N-1} \int_{t_n}^{t_{n+1}} \mathbb{E}\Big[\tau \int_{t_n}^{t_{n+1}} \|\Delta_h X_h(s)\|^2_{\dot{\mathbb{H}}^\gamma_h} + \|f_h(s)\|^2_{\dot{\mathbb{H}}^\gamma_h}\,\mathrm{d}s \\
&\qquad\quad + \int_{t_n}^{t_{n+1}} \|X_h(s)\|^2_{\dot{\mathbb{H}}^\gamma_h} + \|g_h(s)\|^2_{\dot{\mathbb{H}}^\gamma_h}\,\mathrm{d}s\Big]\,\mathrm{d}t \\
&\quad \le C\tau\Big\{\|\Pi_h x\|^2_{\dot{\mathbb{H}}^{\gamma+1}_h} + \mathbb{E}\Big[\int_0^T \|f_h(t)\|^2_{\dot{\mathbb{H}}^\gamma_h} + \|g_h(t)\|^2_{\dot{\mathbb{H}}^{\gamma+1}_h}\,\mathrm{d}t\Big]\Big\}\,,
\end{aligned}
$$

which, together with $\|\Pi_h x\|_{\dot{\mathbb{H}}^1_h} \le C\|x\|_{\dot{\mathbb{H}}^1}$ and $\|\Pi_h x\|_{\dot{\mathbb{H}}^2_h} \le C\|x\|_{\dot{\mathbb{H}}^2}$ (see (2.4a), (2.4d) in Lemma 2.2), settles the desired assertion. □

The following result bounds the difference between $X(\cdot)$ and $X_h(\cdot)$.

Theorem 3.1 *Let $X(\cdot)$ resp. $X_h(\cdot)$ be solutions to (2.12) resp. (3.1). Then there exists a constant C such that*

$$\begin{aligned}
&\mathbb{E}\Big[\sup_{t\in[0,T]}\|X(t)-X_h(t)\|^2_{\dot{\mathbb{H}}^{\gamma}}\Big]+\mathbb{E}\Big[\int_0^T\|X(t)-X_h(t)\|^2_{\dot{\mathbb{H}}^{\gamma+1}}\,\mathrm{d}t\Big]\\
&\leq Ch^{2(1-\gamma)}\Big\{\|x\|^2_{\mathbb{H}^1_0}+\mathbb{E}\Big[\int_0^T\|f(t)\|^2+\|g(t)\|^2_{\mathbb{H}^1_0}\,\mathrm{d}t\Big]\Big\}\\
&\quad+C\,\mathbb{E}\Big[\int_0^T\|\Pi_h f(t)-f_h(t)\|^2_{\dot{\mathbb{H}}_h^{\gamma-1}}+\|\Pi_h g(t)-g_h(t)\|^2_{\dot{\mathbb{H}}_h^{\gamma}}\,\mathrm{d}t\Big],
\end{aligned}\tag{3.4}$$

where $\gamma=-1\,,0$, and

$$\begin{aligned}
&\mathbb{E}\Big[\sup_{t\in[0,T]}\|X(t)-X_h(t)\|^2\Big]\\
&\leq Ch^4\Big\{\|x\|^2_{\mathbb{H}^1_0\cap\mathbb{H}^2}+\mathbb{E}\Big[\int_0^T\|f(s)\|^2_{\mathbb{H}^1_0\cap\mathbb{H}^2}+\|g(s)\|^2_{\mathbb{H}^1_0\cap\mathbb{H}^2}\,\mathrm{d}s\Big]\Big\}\\
&\quad+C\,\mathbb{E}\Big[\int_0^T\big\|\Pi_h\big[f(s)-f_h(s)\big]\big\|^2+\big\|\Pi_h\big[g(s)-g_h(s)\big]\big\|^2\,\mathrm{d}s\Big].
\end{aligned}\tag{3.5}$$

Proof (1) Verification of (3.4). Define the $\mathbb{V}_h$-valued process $e_X(\cdot)=\Pi_h X(\cdot)-X_h(\cdot)$. We have

$$\begin{cases}
\mathrm{d}e_X(t)=\big[\Delta_h e_X(t)+(\Pi_h\Delta-\Delta_h\Pi_h)X(t)+\big(\Pi_h f(t)-f_h(t)\big)\big]\,\mathrm{d}t\\
\qquad\qquad+\big[\beta e_X(t)+\big(\Pi_h g(t)-g_h(t)\big)\big]\,\mathrm{d}W(t)\qquad t\in(0,T]\,,\\
e_X(0)=0\,.
\end{cases}$$

Then Lemmata 3.1 and 2.2 yield

$$\begin{aligned}
&\mathbb{E}\Big[\sup_{t\in[0,T]}\|e_X(t)\|^2_{\mathbb{H}^{-1}}\Big]+\mathbb{E}\Big[\int_0^T\|e_X(t)\|^2\,\mathrm{d}t\Big]\\
&\leq C\,\mathbb{E}\Big[\int_0^T\|(\Pi_h\Delta-\Delta_h\Pi_h)X(t)\|^2_{\dot{\mathbb{H}}_h^{-2}}\Big]\\
&\quad+C\,\mathbb{E}\Big[\int_0^T\|\Pi_h f(t)-f_h(t)\|^2_{\dot{\mathbb{H}}_h^{-2}}+\|\Pi_h g(t)-g_h(t)\|^2_{\dot{\mathbb{H}}_h^{-1}}\,\mathrm{d}t\Big]\\
&\leq Ch^4\Big\{\|x\|^2_{\mathbb{H}^1_0}+\mathbb{E}\Big[\int_0^T\|f(t)\|^2+\|g(t)\|^2_{\mathbb{H}^1_0}\,\mathrm{d}t\Big]\Big\}\\
&\quad+C\,\mathbb{E}\Big[\int_0^T\|\Pi_h f(t)-f_h(t)\|^2_{\dot{\mathbb{H}}_h^{-2}}+\|\Pi_h g(t)-g_h(t)\|^2_{\dot{\mathbb{H}}_h^{-1}}\,\mathrm{d}t\Big].
\end{aligned}\tag{3.6}$$

Now, by (3.6) and Lemmata 2.5, 2.2, we conclude that

$$
\begin{aligned}
&\mathbb{E}\Big[\sup_{t\in[0,T]} \|X(t)-X_h(t)\|^2_{\mathbb{H}^{-1}}\Big] + \mathbb{E}\Big[\int_0^T \|X(t)-X_h(t)\|^2\,\mathrm{d}t\Big] \\
&\quad\le C\Big\{\mathbb{E}\Big[\sup_{t\in[0,T]} \|X(t)-\Pi_h X(t)\|^2_{\mathbb{H}^{-1}}\Big] + \mathbb{E}\Big[\sup_{t\in[0,T]} \|e_X(t)\|^2_{\mathbb{H}^{-1}}\Big] \\
&\qquad + \mathbb{E}\Big[\int_0^T \|X(t)-\Pi_h X(t)\|^2\,\mathrm{d}t\Big] + \mathbb{E}\Big[\int_0^T \|e_X(t)\|^2\,\mathrm{d}t\Big]\Big\} \\
&\quad\le Ch^4\Big\{\|x\|^2_{\mathbb{H}^1_0} + \mathbb{E}\Big[\int_0^T \|f(t)\|^2 + \|g(t)\|^2_{\mathbb{H}^1_0}\,\mathrm{d}t\Big]\Big\} \\
&\qquad + C\,\mathbb{E}\Big[\int_0^T \|\Pi_h f(t)-f_h(t)\|^2_{\dot{\mathbb{H}}_h^{-2}} + \|\Pi_h g(t)-g_h(t)\|^2_{\dot{\mathbb{H}}_h^{-1}}\,\mathrm{d}t\Big].
\end{aligned}
$$

That completes the assertion (3.4) for $\gamma = -1$.

A corresponding argument yields the assertion (3.4) for $\gamma = 0$.

(2) Verification of (3.5). We use tools from semigroup theory. By (2.12) and (3.1), it follows that

$$
\begin{aligned}
X(t) &= E(t)x + \int_0^t E(t-s)f(s)\,\mathrm{d}s + \int_0^t E(t-s)\big[\beta X(s)+g(s)\big]\,\mathrm{d}W(s), \\
X_h(t) &= E_h(t)\Pi_h x + \int_0^t E_h(t-s)f_h(s)\,\mathrm{d}s \\
&\quad + \int_0^t E_h(t-s)\big[\beta X_h(s)+g_h(s)\big]\,\mathrm{d}W(s).
\end{aligned}
$$

On setting $\delta X(\cdot) = X(\cdot) - X_h(\cdot)$ and using the notation $G_h(\cdot)$ from (2.6),

$$
\begin{aligned}
\delta X(t) &= G_h(t)x + \int_0^t G_h(t-s)f(s) + E_h(t-s)\Pi_h\big[f(s)-f_h(s)\big]\,\mathrm{d}s \\
&\quad + \int_0^t \beta G_h(t-s)X(s) + \beta E_h(t-s)\delta X(s) \\
&\qquad + G_h(t-s)g(s) + E_h(t-s)\Pi_h\big[g(s)-g_h(s)\big]\,\mathrm{d}W(s).
\end{aligned}
$$

After squaring both sides, taking expectations, applying Lemma 2.3 with $\rho = \gamma = 2$ and Lemma 2.5, we can deduce that

$$
\begin{aligned}
&\mathbb{E}\big[\|\delta X(t)\|^2\big] \\
&\le C\Big\{\|G_h(t)x\|^2 + \mathbb{E}\Big[\int_0^t \|G_h(t-s)f(s)\|^2 + \|G_h(t-s)X(s)\|^2 \\
&\quad + \|G_h(t-s)g(s)\|^2 + \|\delta X(s)\|^2 + \big\|\Pi_h\big[f(s)-f_h(s)\big]\big\|^2
\end{aligned}
$$

$$
\begin{aligned}
&+\big\|\Pi_h\big[g(s)-g_h(s)\big]\big\|^2\,\mathrm{d}s\Big]\Big\}\\
&\le Ch^4\Big\{\|x\|^2_{\mathbb{H}_0^1\cap\mathbb{H}^2}+\mathbb{E}\Big[\int_0^t\|f(s)\|^2_{\mathbb{H}_0^1\cap\mathbb{H}^2}+\|g(s)\|^2_{\mathbb{H}_0^1\cap\mathbb{H}^2}\,\mathrm{d}s\Big]\Big\}\\
&+C\mathbb{E}\Big[\int_0^t\|\delta X(s)\|^2+\big\|\Pi_h\big[f(s)-f_h(s)\big]\big\|^2+\big\|\Pi_h\big[g(s)-g_h(s)\big]\big\|^2\,\mathrm{d}s\Big],
\end{aligned}
$$

which, together with Gronwall's inequality, yields the assertion with the supremum outside $\mathbb{E}[\cdot]$; using the BDG inequality then implies (3.5). □

3.1.2 *Temporal Discretization and Error Estimates for Linear SDEs*

In this part, we adopt the Euler method to temporally discretize (3.1), i.e.,

$$
\begin{cases}
X_{n+1}-X_n=\tau\big[\Delta_h X_{n+1}+\widetilde{f}_h(t_n)\big]+\big[\beta X_n+\widetilde{g}_h(t_n)\big]\Delta_{n+1}W & n=0,1,\cdots,N,\\
x=\Pi_h x\,,
\end{cases}
\tag{3.7}
$$

where $\widetilde{f}_h(\cdot)\,,\widetilde{g}_h(\cdot)\in D_{\mathbb{F}}\big([0,T];L^2(\Omega;\mathbb{V}_h)\big)$ are approximations of $f_h(\cdot)\,,g_h(\cdot)$ respectively. While the next lemma settles relevant stability properties for (3.7), Theorem 3.2 provides rates for it.

Lemma 3.2 *Let* $X_\cdot\equiv\{X_n\}_{n=0}^N$ *be the solution of* (3.7)*. There exists a constant* C *such that*

$$
\begin{aligned}
&\max_{0\le n\le N}\mathbb{E}\big[\|X_n\|^2_{\dot{\mathbb{H}}_h^\gamma}\big]+\sum_{n=0}^{N-1}\mathbb{E}\big[\|X_{n+1}-X_n\|^2_{\dot{\mathbb{H}}_h^\gamma}\big]+\tau\sum_{n=1}^{N}\mathbb{E}\big[\|X_n\|^2_{\dot{\mathbb{H}}_h^{\gamma+1}}\big]\\
&\quad\le C\Big[\|x\|^2_{\dot{\mathbb{H}}_h^\gamma}+\max_{0\le n\le N}\mathbb{E}\big[\|\widetilde{f}_h(t_n)\|^2_{\dot{\mathbb{H}}_h^{\gamma-1}}+\|\widetilde{g}_h(t_n)\|^2_{\dot{\mathbb{H}}_h^\gamma}\big]\Big]\qquad\gamma=0,1\,.
\end{aligned}
$$

Proof Here we only prove the case $\gamma=0$; the case $\gamma=1$ can be derived accordingly. By testing (3.7) with X_{n+1}, using the relation $(x-y,x)=\frac{1}{2}(\|x\|^2-\|y\|^2+\|x-y\|^2)$, and then taking expectations, we arrive at

$$
\begin{aligned}
&\frac{1}{2}\Big\{\mathbb{E}\big[\|X_{n+1}\|^2\big]-\mathbb{E}\big[\|X_n\|^2\big]+\mathbb{E}\big[\|X_{n+1}-X_n\|^2\big]\Big\}+\tau\mathbb{E}\big[\|X_{n+1}\|^2_{\dot{\mathbb{H}}_h^1}\big]\\
&\quad=\tau\mathbb{E}\big[\big(\widetilde{f}_h(t_n),X_{n+1}\big)_{\mathbb{L}^2}\big]+\mathbb{E}\big[\big(\Delta_{n+1}W\big[\beta X_n+\widetilde{g}_h(t_n)\big],X_{n+1}-X_n\big)_{\mathbb{L}^2}\big],
\end{aligned}
$$

since $-\mathbb{E}\big[\big(\Delta_{n+1}W\big[\beta X_n+\widetilde{g}_h(t_n)\big],X_n\big)_{\mathbb{L}^2}\big]=0$. By Young's inequality and the Itô isometry, we resume that

$$\leq \frac{\tau}{2}\mathbb{E}\big[\|X_{n+1}\|^2_{\dot{\mathbb{H}}^1_h}\big] + \frac{\tau}{2}\mathbb{E}\big[\|\widetilde{f}_h(t_n)\|^2_{\dot{\mathbb{H}}^{-1}_h}\big] + \frac{1}{4}\mathbb{E}\big[\|X_{n+1} - X_n\|^2\big] + 2\beta^2\tau\mathbb{E}\big[\|X_n\|^2\big] + 2\tau\mathbb{E}\big[\|\widetilde{g}_h(t_n)\|^2\big].$$

And then

$$\begin{aligned} &\mathbb{E}\big[\|X_{n+1}\|^2\big] + \frac{1}{2}\mathbb{E}\big[\|X_{n+1} - X_n\|^2\big] + \tau\mathbb{E}\big[\|X_{n+1}\|^2_{\dot{\mathbb{H}}^1_h}\big] \\ &\leq \big[1 + 4\beta^2\tau\big]\mathbb{E}\big[\|X_n\|^2\big] + \tau\mathbb{E}\big[\|\widetilde{f}_h(t_n)\|^2_{\dot{\mathbb{H}}^{-1}_h}\big] + 4\tau\mathbb{E}\big[\|\widetilde{g}_h(t_n)\|^2\big]. \end{aligned} \tag{3.8}$$

By applying discrete Gronwall's inequality (see e.g. [1, Lemma A.3]) and then taking the summation in (3.8) over n from 0 to $N-1$, we get

$$\begin{aligned} &\mathbb{E}\big[\|X_N\|^2\big] + \frac{1}{2}\sum_{n=0}^{N-1}\mathbb{E}\big[\|X_{n+1} - X_n\|^2\big] + \tau\sum_{n=0}^{N-1}\mathbb{E}\big[\|X_{n+1}\|^2_{\dot{\mathbb{H}}^1_h}\big] \\ &\leq C\Big[\|x\|^2 + \max_{0\leq n\leq N}\mathbb{E}\big[\|\widetilde{f}_h(t_n)\|^2_{\dot{\mathbb{H}}^{-1}_h}\big] + \max_{0\leq n\leq N}\mathbb{E}\big[\|\widetilde{g}_h(t_n)\|^2\big]\Big]. \end{aligned}$$

That completes the proof. □

Theorem 3.2 *Let $X_h(\cdot)$ be the solution to SDE* (3.1), *and $X_\cdot$ solves* (3.7). *There exists a constant C such that*

$$\begin{aligned} &\max_{0\leq n\leq N}\mathbb{E}\big[\|X_h(t_n) - X_n\|^2_{\dot{\mathbb{H}}^{-1}_h}\big] + \sum_{n=0}^{N-1}\mathbb{E}\Big[\int_{t_n}^{t_{n+1}}\|X_h(t) - X_n\|^2\,\mathrm{d}t\Big] \\ &\leq C\tau\Big\{\|x\|^2_{\mathbb{H}^1_0} + \mathbb{E}\Big[\int_0^T\|f_h(t)\|^2 + \|g_h(t)\|^2_{\mathbb{H}^1_0}\,\mathrm{d}t\Big]\Big\} \\ &\quad + C\sum_{n=0}^{N-1}\mathbb{E}\Big[\int_{t_n}^{t_{n+1}}\|f_h(t) - \widetilde{f}_h(t_n)\|^2_{\dot{\mathbb{H}}^{-1}_h} + \|g_h(t) - \widetilde{g}_h(t_n)\|^2_{\dot{\mathbb{H}}^{-1}_h}\,\mathrm{d}t\Big], \end{aligned} \tag{3.9}$$

and

$$\begin{aligned} &\max_{0\leq n\leq N-1}\sup_{t\in[t_n,t_{n+1})}\mathbb{E}\big[\|X_h(t) - X_n\|^2\big] + \sum_{n=0}^{N-1}\mathbb{E}\Big[\int_{t_n}^{t_{n+1}}\|X_h(t) - X_n\|^2_{\mathbb{H}^1_0}\,\mathrm{d}t\Big] \\ &\leq C\tau\Big\{\|x\|^2_{\mathbb{H}^1_0\cap\mathbb{H}^2} + \mathbb{E}\Big[\int_0^T\|f_h(t)\|^2_{\mathbb{H}^1_0} + \|g_h(t)\|^2_{\mathbb{H}^1_0\cap\mathbb{H}^2}\,\mathrm{d}t\Big]\Big\} \\ &\quad + C\sum_{n=0}^{N-1}\mathbb{E}\Big[\int_{t_n}^{t_{n+1}}\|f_h(t) - \widetilde{f}_h(t_n)\|^2 + \|g_h(t) - \widetilde{g}_h(t_n)\|^2\,\mathrm{d}t\Big]. \end{aligned} \tag{3.10}$$

Proof We only prove (3.9). The argumentation for (3.10) is very close to (3.9) and we again leave it to the interested reader.

For any $n = 0, 1, \cdots, N$, consider the $\mathbb{V}_h$-valued random variable $e_n = X_n - X_h(t_n)$. Then $e_\cdot$ satisfies

$$\begin{cases} e_{n+1} - e_n = \tau\Delta_h e_{n+1} + \displaystyle\int_{t_n}^{t_{n+1}} \big[\Delta_h(X_h(t_{n+1}) - X_h(t))\big] + \big[\widetilde{f}_h(t_n) - f_h(t)\big]\mathrm{d}t \\ \qquad + \beta e_n \Delta_{n+1} W + \displaystyle\int_{t_n}^{t_{n+1}} \beta\big[X_h(t_n) - X_h(t)\big] + \big[\widetilde{g}_h(t_n) - g_h(t)\big]\mathrm{d}W(t) \\ \qquad\qquad n = 0, 1, \cdots, N-1, \\ e_0 = 0. \end{cases} \tag{3.11}$$

By testing this equation with $(-\Delta_h)^{-1} e_{n+1}$ and using the same idea as in the proof of Lemma 3.2, we have

$$\begin{aligned} &\frac{1}{2}\Big\{\mathbb{E}\big[\|(-\Delta_h)^{-1/2} e_{n+1}\|^2\big] - \mathbb{E}\big[\|(-\Delta_h)^{-1/2} e_n\|^2\big] \\ &\quad + \mathbb{E}\big[\|(-\Delta_h)^{-1/2}(e_{n+1} - e_n)\|^2\big]\Big\} + \tau\mathbb{E}\big[\|e_{n+1}\|^2\big] \\ &= \mathbb{E}\Big[\int_{t_n}^{t_{n+1}} \big(X_h(t) - X_h(t_{n+1}), e_{n+1}\big)_{\mathbb{L}^2}\,\mathrm{d}t\Big] \\ &\quad + \mathbb{E}\Big[\int_{t_n}^{t_{n+1}} \big((-\Delta_h)^{-1/2}\big(\widetilde{f}_h(t_n) - f_h(t)\big), (-\Delta_h)^{-1/2}(e_{n+1} - e_n)\big)_{\mathbb{L}^2}\,\mathrm{d}t\Big] \\ &\quad + \mathbb{E}\Big[\int_{t_n}^{t_{n+1}} \big((-\Delta_h)^{-1/2}\big(\widetilde{f}_h(t_n) - f_h(t)\big), (-\Delta_h)^{-1/2} e_n\big)_{\mathbb{L}^2}\,\mathrm{d}t\Big] \\ &\quad + \mathbb{E}\Big[\big(\Delta_{n+1} W \beta(-\Delta_h)^{-1/2} e_n, (-\Delta_h)^{-1/2}(e_{n+1} - e_n)\big)_{\mathbb{L}^2}\Big] \\ &\quad + \mathbb{E}\Big[\Big(\int_{t_n}^{t_{n+1}} (-\Delta_h)^{-1/2}\beta\big(X_h(t_n) - X_h(t)\big)\,\mathrm{d}W(t), (-\Delta_h)^{-1/2}(e_{n+1} - e_n)\Big)_{\mathbb{L}^2}\Big] \\ &\quad + \mathbb{E}\Big[\Big(\int_{t_n}^{t_{n+1}} (-\Delta_h)^{-1/2}\big(\widetilde{g}_h(t_n) - g_h(t)\big)\mathrm{d}W(t), (-\Delta_h)^{-1/2}(e_{n+1} - e_n)\Big)_{\mathbb{L}^2}\Big]. \end{aligned} \tag{3.12}$$

A standard procedure, which uses the Itô isometry, Young's inequality, discrete Gronwall's inequality and the fact that $e_0 = 0$, then leads to

$$\begin{aligned} &\max_{0 \le k \le N} \mathbb{E}\big[\|e_k\|^2_{\dot{\mathbb{H}}_h^{-1}}\big] \\ &\quad \le C\sum_{n=0}^{N-1}\Big\{\mathbb{E}\Big[\int_{t_n}^{t_{n+1}} \|X_h(t) - X_h(t_{n+1})\|^2 + \|X_h(t) - X_h(t_n)\|^2_{\dot{\mathbb{H}}_h^{-1}}\,\mathrm{d}t\Big] \\ &\quad + \mathbb{E}\Big[\int_{t_n}^{t_{n+1}} \|f_h(t) - \widetilde{f}_h(t_n)\|^2_{\dot{\mathbb{H}}_h^{-1}} + \|g_h(t) - \widetilde{g}_h(t_n)\|^2_{\dot{\mathbb{H}}_h^{-1}}\,\mathrm{d}t\Big]\Big\}. \end{aligned}$$

We may now use this result to verify assertion (3.9). Therefore, we come back to (3.12) and do summation over n from 0 to $N-1$ to conclude

$$
\begin{aligned}
&\sum_{n=0}^{N-1}\Big[\mathbb{E}\big[\|e_{n+1}-e_n\|^2_{\dot{\mathbb{H}}_h^{-1}}\big]+\tau\mathbb{E}\big[\|e_{n+1}\|^2\big]\Big] \\
&\quad\le (1+4\beta^2)\tau\sum_{n=0}^{N-1}\mathbb{E}\big[\|e_n\|^2_{\dot{\mathbb{H}}_h^{-1}}\big] \\
&\qquad+\sum_{n=0}^{N-1}\Big\{\mathbb{E}\Big[\int_{t_n}^{t_{n+1}}\|X_h(t)-X_h(t_{n+1})\|^2+8\beta^2\|X_h(t)-X_h(t_n)\|^2_{\dot{\mathbb{H}}_h^{-1}}\,\mathrm{d}t\Big] \\
&\qquad+\mathbb{E}\Big[\int_{t_n}^{t_{n+1}}(4\tau+1)\|f_h(t)-\widetilde{f}_h(t_n)\|^2_{\dot{\mathbb{H}}_h^{-1}}+8\|g_h(t)-\widetilde{g}_h(t_n)\|^2_{\dot{\mathbb{H}}_h^{-1}}\,\mathrm{d}t\Big]\Big\} \\
&\quad\le C\sum_{n=0}^{N-1}\Big\{\mathbb{E}\Big[\int_{t_n}^{t_{n+1}}\|X_h(t)-X_h(t_{n+1})\|^2+\|X_h(t)-X_h(t_n)\|^2_{\dot{\mathbb{H}}_h^{-1}}\,\mathrm{d}t\Big] \\
&\qquad+\mathbb{E}\Big[\int_{t_n}^{t_{n+1}}\|f_h(t)-\widetilde{f}_h(t_n)\|^2_{\dot{\mathbb{H}}_h^{-1}}+\|g_h(t)-\widetilde{g}_h(t_n)\|^2_{\dot{\mathbb{H}}_h^{-1}}\,\mathrm{d}t\Big]\Big\}.
\end{aligned}
$$

This bound, together with (3.3) for $\gamma = 0$ in Lemma 3.1, then leads to the estimate

$$
\begin{aligned}
&\max_{0\le n\le N}\mathbb{E}\big[\|e_n\|^2_{\dot{\mathbb{H}}_h^{-1}}\big]+\sum_{n=0}^{N-1}\mathbb{E}\Big[\int_{t_n}^{t_{n+1}}\|X_h(t)-X_n\|^2\,\mathrm{d}t\Big] \\
&\quad\le \max_{0\le n\le N}\mathbb{E}\big[\|e_n\|^2_{\dot{\mathbb{H}}_h^{-1}}\big]+C\sum_{n=0}^{N-1}\mathbb{E}\Big[\int_{t_n}^{t_{n+1}}\|X_h(t)-X_h(t_n)\|^2\,\mathrm{d}t+\tau\|e_{n+1}\|^2\Big] \\
&\quad\le C\tau\Big\{\|x\|^2_{\mathbb{H}_0^1}+\mathbb{E}\Big[\int_0^T\|f_h(t)\|^2+\|g_h(t)\|^2_{\mathbb{H}_0^1}\,\mathrm{d}t\Big]\Big\} \\
&\qquad+C\sum_{n=0}^{N-1}\mathbb{E}\Big[\int_{t_n}^{t_{n+1}}\|f_h(t)-\widetilde{f}_h(t_n)\|^2_{\dot{\mathbb{H}}_h^{-1}}+\|g_h(t)-\widetilde{g}_h(t_n)\|^2_{\dot{\mathbb{H}}_h^{-1}}\,\mathrm{d}t\Big].
\end{aligned}
$$

That settles the assertion (3.9). □

Remark 3.1 For SDE (3.1) with additive noise, i.e., $\beta = 0$, the order of convergence for the Euler method could be higher; see e.g. [2, 3].

3.2 Discretization of Linear BSPDEs and Rates

We split the derivation of rates of convergence for the scheme (3.19) to discretize BSPDE (2.16) into two steps: in Sect. 3.2.1 we begin with a backward stochastic differential equation, abbreviated as BSDE, given in (3.13), which serves as a spatial discretization of (2.16); related stability bounds for the solutions to BSDEs may then serve in Sect. 3.2.2 to verify rates of convergence for the complete discretization in space and time.

3.2.1 Spatial Discretization and Error Estimates for Linear BSPDEs

We now consider a finite element method of the BSPDE (2.16). Let $Y_{T,h} \in L^2_{\mathcal{F}_T}(\Omega; \mathbb{V}_h)$, $f_h(\cdot) \in L^2_{\mathbb{F}}(0, T; \mathbb{V}_h)$ be approximations of Y_T, $f(\cdot)$, respectively. Then the finite element discretization of (2.16) is the following BSDE:

$$\begin{cases} \mathrm{d}Y_h(t) = \big[-\Delta_h Y_h(t) - \beta Z_h(t) + f_h(t)\big]\,\mathrm{d}t + Z_h(t)\,\mathrm{d}W(t) \quad t \in [0, T)\,, \\ Y_h(T) = Y_{T,h}\,. \end{cases} \tag{3.13}$$

The well-posedness of a solution $\big(Y_h(\cdot), Z_h(\cdot)\big) \in L^2_{\mathbb{F}}\big(\Omega; C([0, T]; \mathbb{V}_h)\big) \times L^2_{\mathbb{F}}(0, T; \mathbb{V}_h)$ follows from [4, Theorem 2.1]. The following result is on the stability of the solution to (3.13).

Lemma 3.3 *Suppose that $Y_{T,h} \in L^2_{\mathcal{F}_T}(\Omega; \mathbb{V}_h)$ and $f_h(\cdot) \in L^2_{\mathbb{F}}(0, T; \mathbb{V}_h)$. There exists a constant C independent of h such that*

$$\begin{aligned} &\mathbb{E}\Big[\sup_{t\in[0,T]} \|Y_h(t)\|^2_{\dot{\mathbb{H}}^\gamma_h}\Big] + \mathbb{E}\Big[\int_0^T \|Y_h(t)\|^2_{\dot{\mathbb{H}}^{\gamma+1}_h} + \|Z_h(t)\|^2_{\dot{\mathbb{H}}^\gamma_h}\,\mathrm{d}t\Big] \\ &\quad \le C\,\mathbb{E}\Big[\|Y_{T,h}\|^2_{\dot{\mathbb{H}}^\gamma_h} + \int_0^T \|f_h(t)\|^2_{\dot{\mathbb{H}}^{\gamma-1}_h}\,\mathrm{d}t\Big] \qquad \gamma = -1, 0, 1, 2\,. \end{aligned} \tag{3.14}$$

For a uniform partition $I_\tau = \{t_n\}_{n=0}^N$ of size $\tau > 0$ covering $[0, T]$, it holds that

$$\begin{aligned} &\sum_{n=0}^{N-1} \mathbb{E}\Big[\int_{t_n}^{t_{n+1}} \|Y_h(t) - Y_h(t_n)\|^2_{\dot{\mathbb{H}}^\gamma_h}\,\mathrm{d}t\Big] \\ &\quad \le C\tau\,\mathbb{E}\Big[\|Y_{T,h}\|^2_{\dot{\mathbb{H}}^{\gamma+1}_h} + \int_0^T \|f_h(t)\|^2_{\dot{\mathbb{H}}^\gamma_h}\,\mathrm{d}t\Big] \qquad \gamma = 0, 1\,. \end{aligned} \tag{3.15}$$

Proof The assertion (3.14) can be derived by Itô's formula, Gronwall's inequality and the BDG inequality; see e.g. [5, Lemma 3.1]. To prove (3.15), we follow a similar technique as in the proof of (3.3) in Lemma 3.1. By the Itô isometry and (3.14), BSDE (3.13) easily gives

$$\begin{aligned} &\sum_{n=0}^{N-1} \mathbb{E}\Big[\int_{t_n}^{t_{n+1}} \|Y_h(t) - Y_h(t_n)\|^2_{\dot{\mathbb{H}}^\gamma_h}\,\mathrm{d}t\Big] \\ &\quad \le C\tau\,\mathbb{E}\Big[\int_0^T \|\Delta_h Y_h(t)\|^2_{\dot{\mathbb{H}}^\gamma_h} + \|Z_h(t)\|^2_{\dot{\mathbb{H}}^\gamma_h} + \|f_h(t)\|^2_{\dot{\mathbb{H}}^\gamma_h}\,\mathrm{d}t\Big] \\ &\quad \le C\tau\,\mathbb{E}\Big[\|Y_{T,h}\|^2_{\dot{\mathbb{H}}^{\gamma+1}_h} + \int_0^T \|f_h(t)\|^2_{\dot{\mathbb{H}}^\gamma_h}\,\mathrm{d}t\Big]. \end{aligned}$$

That completes the proof. □

The following result settles rates of convergence for discretization (3.13), which also improves the result in [5, Theorem 3.2].

Theorem 3.3 *Suppose that* $Y_T \in L^2_{\mathcal{F}_T}(\Omega; \mathbb{H}^1_0 \cap \mathbb{H}^2)$ *and* $f(\cdot) \in L^2_{\mathbb{F}}(0, T; \mathbb{H}^1_0 \cap \mathbb{H}^2)$. *Let* $\big(Y(\cdot), Z(\cdot)\big)$ *resp.* $\big(Y_h(\cdot), Z_h(\cdot)\big)$ *be solutions of* (2.16) *resp.* (3.13). *There exists a constant* C *independent of* h *such that*

$$\begin{aligned}
&\mathbb{E}\Big[\sup_{t\in[0,T]} \|Y(t)-Y_h(t)\|^2_{\dot{\mathbb{H}}^\gamma}\Big]+\mathbb{E}\Big[\int_0^T \|Y(t)-Y_h(t)\|^2_{\dot{\mathbb{H}}^{\gamma+1}}+\|Z(t)-Z_h(t)\|^2_{\dot{\mathbb{H}}^\gamma}\,\mathrm{d}t\Big]\\
&\le Ch^{2(1-\gamma)}\mathbb{E}\Big[\|Y_T\|^2_{\mathbb{H}^1_0}+\int_0^T \|f(t)\|^2\,\mathrm{d}t\Big]\\
&\quad +C\,\mathbb{E}\Big[\|\Pi_h Y_T - Y_{T,h}\|^2_{\dot{\mathbb{H}}^\gamma_h}+\int_0^T \|\Pi_h f(t)-f_h(t)\|^2_{\dot{\mathbb{H}}^{\gamma-1}_h}\,\mathrm{d}t\Big] \quad \gamma=-1,0, \quad (3.16)
\end{aligned}$$

and

$$\begin{aligned}
&\sup_{t\in[0,T]} \mathbb{E}\big[\|Y(t)-Y_h(t)\|^2\big]\\
&\qquad\le Ch^4\mathbb{E}\Big[\|Y_T\|^2_{\mathbb{H}^1_0\cap\mathbb{H}^2}+\int_0^T \|f(t)\|^2_{\mathbb{H}^1_0\cap\mathbb{H}^2}\,\mathrm{d}t\Big]\\
&\qquad\quad +C\mathbb{E}\Big[\|Y_T-Y_{T,h}\|^2+\int_0^T \|f(t)-f_h(t)\|^2\,\mathrm{d}t\Big]. \qquad (3.17)
\end{aligned}$$

Proof (1) Verification of (3.16). By defining two $\mathbb{V}_h$-valued stochastic processes $e_Y(\cdot) = \Pi_h Y(\cdot) - Y_h(\cdot)$ and $e_Z(\cdot) = \Pi_h Z(\cdot) - Z_h(\cdot)$, we find that

$$\begin{cases}
\mathrm{d}e_Y(t) = \big[-\Delta_h e_Y(t) - \beta e_Z(t) - (\Pi_h\Delta - \Delta_h\Pi_h)Y(t)\\
\qquad\qquad +\big(\Pi_h f(t) - f_h(t)\big)\big]\,\mathrm{d}t + e_Z(t)\,\mathrm{d}W(t) \qquad t\in[0,T),\\
e_Y(T) = \Pi_h Y_T - Y_{T,h}\,.
\end{cases}$$

Based on (3.14) in Lemma 3.3, (2.4c) in Lemma 2.2, and Lemma 2.6, we have

$$\begin{aligned}
&\mathbb{E}\Big[\sup_{t\in[0,T]} \|e_Y(t)\|^2_{\dot{\mathbb{H}}^{-1}_h}\Big]+\mathbb{E}\Big[\int_0^T \|e_Y(t)\|^2+\|e_Z(t)\|^2_{\dot{\mathbb{H}}^{-1}_h}\,\mathrm{d}t\Big]\\
&\le C\,\mathbb{E}\Big[\|\Pi_h Y_T - Y_{T,h}\|^2_{\dot{\mathbb{H}}^{-1}_h}+\int_0^T \|(-\Delta_h)^{-1}(\Delta_h\mathcal{R}_h - \Delta_h\Pi_h)Y(t)\|^2\\
&\quad +\|\Pi_h f(t)-f_h(t)\|^2_{\dot{\mathbb{H}}^{-2}_h}\,\mathrm{d}t\Big]\\
&\le Ch^4\mathbb{E}\Big[\|Y_T\|^2_{\mathbb{H}^1_0}+\int_0^T \|f(t)\|^2\,\mathrm{d}t\Big]\\
&\quad +C\,\mathbb{E}\Big[\|\Pi_h Y_T - Y_{T,h}\|^2_{\dot{\mathbb{H}}^{-1}_h}+\int_0^T \|\Pi_h f(t)-f_h(t)\|^2_{\dot{\mathbb{H}}^{-2}_h}\,\mathrm{d}t\Big].
\end{aligned}$$

On the other hand, Lemmata 2.2 and 2.6 imply that

$$
\begin{aligned}
&\mathbb{E}\Big[\sup_{t\in[0,T]} \|Y(t) - \Pi_h Y(t)\|^2_{\dot{\mathbb{H}}_h^{-1}}\Big] \\
&\qquad + \mathbb{E}\Big[\int_0^T \|Y(t) - \Pi_h Y(t)\|^2 + \|Z(t) - \Pi_h Z(t)\|^2_{\dot{\mathbb{H}}_h^{-1}}\,\mathrm{d}t\Big] \\
&\leq Ch^4\Big\{\sup_{t\in[0,T]} \mathbb{E}\big[\|Y(t)\|^2_{\mathbb{H}_0^1}\big] + \mathbb{E}\Big[\int_0^T \|Y(t)\|^2_{\mathbb{H}_0^1\cap\mathbb{H}^2} + \|Z(t)\|^2_{\mathbb{H}_0^1}\,\mathrm{d}t\Big]\Big\} \\
&\leq Ch^4\Big\{\mathbb{E}\big[\|Y_T\|^2_{\mathbb{H}_0^1}\big] + \mathbb{E}\Big[\int_0^T \|f(t)\|^2\,\mathrm{d}t\Big]\Big\}.
\end{aligned}
$$

The assertion for $\gamma = -1$ then follows from the estimates above.

To derive assertion (3.16) for $\gamma = 0$, we may proceed correspondingly. Firstly by applying (3.14) in Lemma 3.3 and (2.4e) in Lemma 2.2, we conclude that

$$
\begin{aligned}
&\mathbb{E}\Big[\sup_{t\in[0,T]} \|e_Y(t)\|^2\Big] + \mathbb{E}\Big[\int_0^T \|\nabla e_Y(t)\|^2 + \|e_Z(t)\|^2\,\mathrm{d}t\Big] \\
&\leq C\,\mathbb{E}\Big[\|\Pi_h Y_T - Y_{T,h}\|^2 \\
&\qquad + \int_0^T \|(\Pi_h\Delta - \Delta_h\Pi_h)Y(t)\|^2_{\dot{\mathbb{H}}_h^{-1}} + \|\Pi_h f(t) - f_h(t)\|^2_{\dot{\mathbb{H}}_h^{-1}}\,\mathrm{d}t\Big] \\
&\leq Ch^2\,\mathbb{E}\Big[\|Y_T\|^2_{\mathbb{H}_0^1} + \int_0^T \|f(t)\|^2\,\mathrm{d}t\Big] \\
&\qquad + C\,\mathbb{E}\Big[\|\Pi_h Y_T - Y_{T,h}\|^2 + \int_0^T \|\Pi_h f(t) - f_h(t)\|^2_{\dot{\mathbb{H}}_h^{-1}}\,\mathrm{d}t\Big].
\end{aligned}
$$

It is now immediate to complete the argument, which settles the assertion for $\gamma = 0$.

(2) Verification of (3.17). To derive the assertion, we consider a time partition $I_{\tau_0} = \{T_n\}_{n=0}^{N_0}$ with $\tau_0 \leq \frac{1}{3\beta^2}$ to control the term led by β; if $\beta = 0$, we do not need any partition. By setting $\delta Y(\cdot) = Y(\cdot) - Y_h(\cdot)$, $\delta f(\cdot) = f(\cdot) - f_h(\cdot)$ and applying (2.16) and (3.13), for any $t \in [T_n, T_{n+1})$ we have

$$
\begin{aligned}
&\delta Y(t) + \int_t^{T_{n+1}} E(s-t)Z(s) - E_h(s-t)Z_h(s)\,\mathrm{d}W(s) \\
&\quad = E(T_{n+1}-t)Y(T_{n+1}) - E_h(T_{n+1}-t)Y_h(T_{n+1}) \\
&\qquad + \beta\int_t^{T_{n+1}} E(s-t)Z(s) - E_h(s-t)Z_h(s)\,\mathrm{d}s \\
&\qquad - \int_t^{T_{n+1}} E(s-t)f(s) - E_h(s-t)f_h(s)\,\mathrm{d}s\,.
\end{aligned}
$$

Using mutual independence of $\delta Y(t)$ and $\int_t^{T_{n+1}} E(s-t)Z(s) - E_h(s-t) Z_h(s)\,\mathrm{d}W(s)$, squaring on both sides, taking expectations, and applying (2.6), we arrive at

$$\begin{aligned}
&\mathbb{E}\big[\|\delta Y(t)\|^2\big] + \mathbb{E}\Big[\int_t^{T_{n+1}} \|E(s-t)Z(s) - E_h(s-t)Z_h(s)\|^2\,\mathrm{d}s\Big] \\
&\quad \le 6\mathbb{E}\big[\|G_h(T_{n+1}-t)Y(T_{n+1})\|^2\big] + 6\mathbb{E}\big[\|E_h(T_{n+1}-t)\Pi_h\delta Y(T_{n+1})\|^2\big] \\
&\quad\quad + 3\beta^2\tau_0\mathbb{E}\Big[\int_t^{T_{n+1}} \|E(s-t)Z(s) - E_h(s-t)Z_h(s)\|^2\,\mathrm{d}s\Big] \\
&\quad\quad + 6\tau_0\mathbb{E}\Big[\int_t^{T_{n+1}} \|G_h(s-t)f(s)\|^2 + \|E_h(s-t)\Pi_h\delta f(s)\|^2\,\mathrm{d}s\Big],
\end{aligned}$$

which, together with the fact that $3\beta^2\tau_0 \le 1$ and Lemma 2.3, yields

$$\begin{aligned}
&\mathbb{E}\big[\|\delta Y(t)\|^2\big] \\
&\le 6\mathbb{E}\big[\|G_h(T_{n+1}-t)Y(T_{n+1})\|^2\big] + 6\mathbb{E}\big[\|E_h(T_{n+1}-t)\Pi_h\delta Y(T_{n+1})\|^2\big] \\
&\quad + 6\tau_0\mathbb{E}\Big[\int_t^{T_{n+1}} \|G_h(s-t)f(s)\|^2 + \|E_h(s-t)\Pi_h\delta f(s)\|^2\,\mathrm{d}s\Big] \\
&\le Ch^4\mathbb{E}\Big[\|Y(T_{n+1})\|^2_{\mathbb{H}_0^1\cap\mathbb{H}^2} + \tau_0\int_{T_n}^{T_{n+1}} \|f(t)\|^2_{\mathbb{H}_0^1\cap\mathbb{H}^2}\,\mathrm{d}t\Big] \\
&\quad + C\,\mathbb{E}\Big[\|\Pi_h\delta Y(T_{n+1})\|^2 + \tau_0\int_{T_n}^{T_{n+1}} \|\Pi_h\delta f(t)\|^2\,\mathrm{d}t\Big].
\end{aligned} \tag{3.18}$$

By taking $t = T_n$, $n = N_0 - 1, N_0 - 2, \cdots, 0$, repeating the above procedure N_0 times and applying (2.17), we conclude that

$$\begin{aligned}
\max_{0\le n\le N_0} \mathbb{E}\big[\|\delta Y(T_n)\|^2\big] &\le Ch^4\mathbb{E}\Big[\|Y(T)\|^2_{\mathbb{H}_0^1\cap\mathbb{H}^2} + \int_0^T \|f(t)\|^2_{\mathbb{H}_0^1\cap\mathbb{H}^2}\,\mathrm{d}t\Big] \\
&\quad + C\,\mathbb{E}\Big[\|\Pi_h\delta Y(T)\|^2 + \int_0^T \|\Pi_h\delta f(t)\|^2\,\mathrm{d}t\Big],
\end{aligned}$$

which, together with (3.18), implies the assertion (3.17). □

3.2.2 *Temporal Discretization and Error Estimates for Linear BSDEs*

In this part, we prove rates of convergence for the temporal discretization of BSDE (3.13) by means of the Euler method:

$$\begin{cases} Y_{h\tau}(t_n) = A_0\mathbb{E}^{t_n}\Big[\big(Y_{h\tau}(t_{n+1})-\tau f_h(t_{n+1})\big)\big(1+\beta\Delta_{n+1}W\big)\Big] & n=0,1,\cdots,N-1\,, \\ Z_{h\tau}(t_n) = \dfrac{1}{\tau}\mathbb{E}^{t_n}\Big[\big(Y_{h\tau}(t_{n+1})-\tau f_h(t_{n+1})\big)\Delta_{n+1}W\Big] & n=0,1,\cdots,N-1, \\ Y_{h\tau}(T) = Y_{T,h}\,, \end{cases} \tag{3.19}$$

where $A_0 = (\mathbb{1}_h - \tau\Delta_h)^{-1}$, and $\mathbb{E}^{t_n}[\cdot]$ is a conditional expectation with respect to $\mathcal{F}_{t_n}$; see part **a.** in Sect. 2.1. To initiate the error analysis, we introduce the following auxiliary equation which serves as a bridge between BSDE (3.13) and difference equation (3.19),

$$\begin{cases} Y_n - Y_{n+1} = \tau\Delta_h Y_n + \displaystyle\int_{t_n}^{t_{n+1}} \beta\bar{Z}_0(t) - \widetilde{f}_h(t)\,\mathrm{d}t \\ \qquad\qquad - \displaystyle\int_{t_n}^{t_{n+1}} Z_0(t)\,\mathrm{d}W(t) \qquad n=0,1,\cdots,N-1, \\ Y_N = \widetilde{Y}_{T,h}\,, \end{cases} \tag{3.20}$$

where

$$\bar{Z}_0(t) = \frac{1}{\tau}\mathbb{E}^{t_n}\Big[\int_{t_n}^{t_{n+1}} Z_0(t)\,\mathrm{d}t\Big] \qquad t\in[t_n,t_{n+1})\,, \tag{3.21}$$

and $\widetilde{Y}_{T,h}$, $\widetilde{f}_h(\cdot)$ are approximations of $Y_{T,h}$, $f_h(\cdot)$ respectively. Note that when $\beta \neq 0$, equation (3.20) is *not implementable*, while (3.19) is, since conditional expectations can be simulated; see Sect. 3.4.2 below.

Remark 3.2 The following observation connects the two equations (3.19) and (3.20). If $\widetilde{Y}_{T,h}$, $\widetilde{f}_h(\cdot)$ are chosen as follows,

$$\widetilde{Y}_{T,h} = Y_{T,h} \qquad \widetilde{f}_h(t) = f_h(t_{n+1}) \qquad t\in[t_n,t_{n+1})\,, n=0,1,\cdots,N-1\,, \tag{3.22}$$

then it holds that

$$Y_{h\tau}(t_n) = Y_n \qquad Z_{h\tau}(t_n) = \bar{Z}_0(t_n) \qquad n=0,1,\cdots,N-1\,.$$

Hence, if an error estimate is obtained for $\big(Y_h(\cdot)-Y_\cdot, Z_h(\cdot)-Z_0(\cdot)\big)$ in a suitable norm, where $\big(Y_h(\cdot), Z_h(\cdot)\big)$ is from (3.13), it transfers to the Euler method (3.19) as well.

The following result is on the well-posedness of (3.20).

Lemma 3.4 *Assume* $\widetilde{Y}_{T,h} \in L^2_{\mathcal{F}_T}(\Omega;\mathbb{V}_h)$ *and* $\widetilde{f}_h(\cdot) \in L^2_{\widetilde{\mathbb{F}}}(0,T;\mathbb{V}_h)$ *with* $\widetilde{\mathbb{F}} = \{\mathcal{F}_{\mu(t)}\}$. *If* $\tau < \frac{1}{\beta^2}$, *then (3.20) admits a unique solution* $\big(Y_\cdot, Z_0(\cdot)\big) \in L^2_{\mathbb{F}}(0,T;\mathbb{V}_h)\times L^2_{\mathbb{F}}(0,T;\mathbb{V}_h)$.

Proof We divide the proof into two steps.

(1) In this step, we consider the case $\beta = 0$. Following the idea to solve BSDEs (see e.g. [6]), for $n = 0, 1, \cdots, N-1$, we define by

$$Y_n = (\mathbb{1}_h - \tau\Delta_h)^{-1}\mathbb{E}^{t_n}\Big[Y_{n+1} - \int_{t_n}^{t_{n+1}} \widetilde{f}_h(t)\,\mathrm{d}t\Big].$$

By the martingale representation theorem, there exists $Z_0(\cdot) \in L^2_{\mathbb{F}}(t_n, t_{n+1}; \mathbb{V}_h)$ such that

$$\int_{t_n}^{t_{n+1}} Z_0(t)\,\mathrm{d}W(t) = (\mathbb{1}_h - \mathbb{E}^{t_n})\Big[Y_{n+1} - \int_{t_n}^{t_{n+1}} \widetilde{f}_h(t)\,\mathrm{d}t\Big].$$

We test (3.20) with Y_n and exploit independence properties to conclude that

$$\begin{aligned}\mathbb{E}\big[(Y_n - Y_{n+1}, Y_n)_{\mathbb{L}^2}\big] = &-\tau\mathbb{E}\big[\big((-\Delta_h)^{1/2}Y_n, (-\Delta_h)^{1/2}Y_n\big)_{\mathbb{L}^2}\big]\\ &- \mathbb{E}\Big[\int_{t_n}^{t_{n+1}} \big\langle(-\Delta_h)^{-1/2}\widetilde{f}_h(t), (-\Delta_h)^{1/2}Y_n\big\rangle\,\mathrm{d}t\Big].\end{aligned}$$

By the binomial formula, we conclude that

$$\mathbb{E}\big[\|Y_n\|^2\big] + \tau\mathbb{E}\big[\|Y_n\|^2_{\dot{\mathbb{H}}^1_h}\big] \le \mathbb{E}\big[\|Y_{n+1}\|^2\big] + \mathbb{E}\Big[\int_{t_n}^{t_{n+1}} \|\widetilde{f}_h(t)\|^2_{\dot{\mathbb{H}}^{-1}_h}\,\mathrm{d}t\Big].$$

For any $k = 0, 1, \cdots, N-1$, after summation over n from k to $N-1$, we may conclude that

$$\max_{0\le k\le N} \mathbb{E}\big[\|Y_k\|^2\big] + \tau\sum_{n=0}^{N-1} \mathbb{E}\big[\|Y_n\|^2_{\dot{\mathbb{H}}^1_h}\big] \le \mathbb{E}\Big[\|\widetilde{Y}_{T,h}\|^2 + \int_0^T \|\widetilde{f}_h(t)\|^2_{\dot{\mathbb{H}}^{-1}_h}\,\mathrm{d}t\Big].$$

Testing (3.20) with $\Delta_h Y_n$ and following the same steps as above now lead to

$$\max_{0\le k\le N} \mathbb{E}\big[\|Y_k\|^2_{\dot{\mathbb{H}}^1_h}\big] + \tau\sum_{n=0}^{N-1} \mathbb{E}\big[\|Y_n\|^2_{\dot{\mathbb{H}}^2_h}\big] \le \mathbb{E}\Big[\|\widetilde{Y}_{T,h}\|^2_{\dot{\mathbb{H}}^1_h} + \int_0^T \|\widetilde{f}_h(t)\|^2\,\mathrm{d}t\Big]. \tag{3.23}$$

By (3.20), it follows that

$$\int_0^T Z_0(t)\,\mathrm{d}W(t) = \widetilde{Y}_{T,h} - Y_0 + \sum_{n=0}^{N-1} \tau\Delta_h Y_n - \int_0^T \widetilde{f}_h(t)\,\mathrm{d}t\,.$$

Then the Itô isometry, binomial formula, and (3.23) lead to

$$\mathbb{E}\Big[\int_0^T \|Z_0(t)\|^2\,\mathrm{d}t\Big] \le C\mathbb{E}\Big[\|\widetilde{Y}_{T,h}\|^2 + \|Y_0\|^2 + \tau\sum_{n=0}^{N-1}\|Y_n\|^2_{\dot{\mathbb{H}}^2_h} + \int_0^T \|\widetilde{f}_h(t)\|^2\,\mathrm{d}t\Big]$$
$$\le C\mathbb{E}\Big[\|\widetilde{Y}_{T,h}\|^2_{\dot{\mathbb{H}}^1_h} + \int_0^T \|\widetilde{f}_h(t)\|^2\,\mathrm{d}t\Big].$$

(2) In this step, we deal with the case $\beta \neq 0$. For any $n = 0, 1, \cdots, N-1$, by step (1), for any $\big(\widehat{Y}_n, \widehat{Z}_0(\cdot)\big) \in L^2_{\mathcal{F}_{t_n}}(\Omega; \mathbb{V}_h) \times L^2_{\mathbb{F}}(t_n, t_{n+1}; \mathbb{V}_h)$ we know that there exists a unique pair $\big(Y_n, Z_0(\cdot)\big) \in L^2_{\mathcal{F}_{t_n}}(\Omega; \mathbb{V}_h) \times L^2_{\mathbb{F}}(t_n, t_{n+1}; \mathbb{V}_h)$ satisfying

$$Y_n - Y_{n+1} = \tau\Delta_h Y_n + \int_{t_n}^{t_{n+1}} \beta\overline{\widehat{Z}_0(t)} - \widetilde{f}_h(t)\,\mathrm{d}t - \int_{t_n}^{t_{n+1}} Z_0(t)\,\mathrm{d}W(t)\,. \tag{3.24}$$

In the following, we show for $\tau < \frac{1}{\beta^2}$ that the map $\mathscr{T}_n : \big(\widehat{Y}_n, \widehat{Z}_0(\cdot)\big) \longmapsto \big(Y_n, Z_0(\cdot)\big)$ defined via (3.24) is a contraction map. To do that, let $\big(\widehat{Y}^i_n, \widehat{Z}^i_0(\cdot)\big) \in L^2_{\mathcal{F}_{t_n}}(\Omega; \mathbb{V}_h) \times L^2_{\mathbb{F}}(t_n, t_{n+1}; \mathbb{V}_h)$ be fixed, for $i = 1, 2$; we then consider $\big(Y^i_n, Z^i_0(\cdot)\big) = \mathscr{T}_n\big((\widehat{Y}^i_n, \widehat{Z}^i_0(\cdot))\big)$. By setting $\big(\delta\widehat{Y}_n, \delta\widehat{Z}_0(\cdot)\big) = \big(\widehat{Y}^1_n - \widehat{Y}^2_n, \widehat{Z}^1_0(\cdot) - \widehat{Z}^2_0(\cdot)\big)$ and $\big(\delta Y_n, s\delta Z_0(\cdot)\big) = \big(Y^1_n - Y^2_n, Z^1_0(\cdot) - Z^2_0(\cdot)\big)$, and exploiting the linear character of the equation (3.20), we have

$$\big(\mathbb{1}_h - \tau\Delta_h\big)\delta Y_n + \int_{t_n}^{t_{n+1}} \delta Z_0(t)\,\mathrm{d}W(t) = \int_{t_n}^{t_{n+1}} \beta\overline{\delta\widehat{Z}_0(t)}\,\mathrm{d}t\,.$$

Since $\mathbb{E}^{t_n}[\delta Y_n] = \delta Y_n$ and $\mathbb{E}^{t_n}\big[\int_{t_n}^{t_{n+1}} \delta Z_0(t)\,\mathrm{d}W(t)\big] = 0$, this implies that

$$\int_{t_n}^{t_{n+1}} \delta Z_0(t)\,\mathrm{d}W(t) = (\mathbb{1}_h - \mathbb{E}^{t_n})\Big[\int_{t_n}^{t_{n+1}} \beta\overline{\delta\widehat{Z}_0(t)}\,\mathrm{d}t\Big],$$
$$\text{and}\quad \delta Y_n = \big(\mathbb{1}_h - \tau\Delta_h\big)^{-1}\mathbb{E}^{t_n}\Big[\int_{t_n}^{t_{n+1}} \beta\overline{\delta\widehat{Z}_0(t)}\,\mathrm{d}t\Big].$$

Then it follows that

$$\mathbb{E}\big[\|\delta Y_n\|^2\big] + \mathbb{E}\Big[\int_{t_n}^{t_{n+1}} \|\delta Z_0(t)\|^2\,\mathrm{d}t\Big] \le \beta^2\tau\mathbb{E}\Big[\int_{t_n}^{t_{n+1}} \|\delta\widehat{Z}_0(t)\|^2\,\mathrm{d}t\Big].$$

Thus, for $\tau < \frac{1}{\beta^2}$, the map $\mathscr{T}_n$ is a contraction; hence by Banach fixed-point theorem, (3.20) admits a unique solution for any $n = 0, 1, \cdots, N-1$. That completes the proof. □

The following result establishes rates of convergence for the temporal discretization (3.20) with $\beta = 0$. The general case will be handled in Theorem 3.5.

Theorem 3.4 *Let* $\big(Y_h(\cdot), Z_h(\cdot)\big)$ *resp.* $\big(Y_\cdot, Z_0(\cdot)\big)$ *be solutions to* (3.13) *resp.* (3.20) *with* $\beta = 0$. *There exists a constant* C *independent of* h *such that*

$$\max_{0\le n\le N} \mathbb{E}\big[\|Y_n - Y_h(t_n)\|^2_{\dot{\mathbb{H}}_h^{-1}}\big] + \sum_{n=0}^{N-1} \tau\mathbb{E}\big[\|Y_n - Y_h(t_n)\|^2\big]$$
$$\le C\tau^2\,\mathbb{E}\Big[\|Y_{T,h}\|^2_{\dot{\mathbb{H}}_h^1} + \int_0^T \|f_h(t)\|^2\,\mathrm{d}t\Big]$$
$$+ C\,\mathbb{E}\Big[\|Y_{T,h} - \widetilde{Y}_{T,h}\|^2_{\dot{\mathbb{H}}_h^{-1}} + \int_0^T \|\widetilde{f}_h(t) - f_h(t)\|^2_{\dot{\mathbb{H}}_h^{-1}}\,\mathrm{d}t\Big]. \tag{3.25}$$

Proof Firstly, we introduce two random equations:

$$\begin{cases} \mathrm{d}\widehat{Y}_h(t) = \big[-\Delta_h\widehat{Y}_h(t) + f_h(t)\big]\mathrm{d}t & t\in[0,T),\\ \widehat{Y}_h(T) = Y_{T,h}, \end{cases} \tag{3.26}$$

and

$$\begin{cases} \widehat{Y}_n - \widehat{Y}_{n+1} = \tau\Delta_h\widehat{Y}_n - \displaystyle\int_{t_n}^{t_{n+1}} \widetilde{f}_h(t)\,\mathrm{d}t & n = 0,1,\cdots,N,\\ \widehat{Y}_N = \widetilde{Y}_{T,h}. \end{cases}$$

(1) We claim that:

$$\begin{aligned} Y_h(t) &= \mathbb{E}^t\big[\widehat{Y}_h(t)\big] & t\in[0,T],\\ \text{and}\quad Y_n &= \mathbb{E}^{t_n}\big[\widehat{Y}_n\big] & n = 0,1,\cdots,N. \end{aligned} \tag{3.27}$$

Indeed, by (3.13) and (3.26), it follows that

$$\begin{aligned} Y_h(t) &= E_h(T-t)Y_{T,h} - \int_t^T E_h(s-t)f_h(s)\,\mathrm{d}s\\ &\quad - \int_t^T E_h(s-t)Z_h(s)\,\mathrm{d}W(s),\\ \text{and}\quad \widehat{Y}_h(t) &= E_h(T-t)Y_{T,h} - \int_t^T E_h(s-t)f_h(s)\,\mathrm{d}s. \end{aligned} \tag{3.28}$$

Hence, since $Y_h(\cdot)$ is $\mathbb{F}$-adapted, we have

$$Y_h(t) = \mathbb{E}^t[Y_h(t)] = \mathbb{E}^t\Big[E_h(T-t)Y_{T,h} - \int_t^T E_h(s-t)f_h(s)\mathrm{d}s\Big] = \mathbb{E}^t\big[\widehat{Y}_h(t)\big],$$

which is the first assertion in (3.27). We use induction to derive the second one. Obviously, $\widehat{Y}_N = Y_N$. Now suppose that $Y_{n+1} = \mathbb{E}^{t_{n+1}}\big[\widehat{Y}_{n+1}\big]$. Similar to (3.28), we can get

$$Y_n = A_0 Y_{n+1} - A_0 \int_{t_n}^{t_{n+1}} \widetilde{f}_h(t)\,\mathrm{d}t - A_0 \int_{t_n}^{t_{n+1}} Z_0(t)\,\mathrm{d}W(t)\,,$$
$$\widehat{Y}_n = A_0 \widehat{Y}_{n+1} - A_0 \int_{t_n}^{t_{n+1}} \widetilde{f}_h(t)\,\mathrm{d}t\,.$$

Then by the tower property for conditional expectations, we deduce that

$$\begin{aligned} Y_n &= \mathbb{E}^{t_n}[Y_n] = \mathbb{E}^{t_n}\Big[A_0 Y_{n+1} - A_0 \int_{t_n}^{t_{n+1}} \widetilde{f}_h(t)\,\mathrm{d}t\Big] \\ &= \mathbb{E}^{t_n}\Big[A_0 \mathbb{E}^{t_{n+1}}\big[\widehat{Y}_{n+1}\big] - A_0 \int_{t_n}^{t_{n+1}} \widetilde{f}_h(t)\,\mathrm{d}t\Big] = \mathbb{E}^{t_n}\big[\widehat{Y}_n\big]\,. \end{aligned}$$

That settles the claim (3.27).

(2) For any $n = 0, 1, \cdots, N$, we define the $\mathbb{V}_h$-valued random variable $e_n = \widehat{Y}_n - \widehat{Y}_h(t_n)$. It then follows that

$$e_n - e_{n+1} = \tau \Delta_h e_n + \int_{t_n}^{t_{n+1}} \Delta_h\big[\widehat{Y}_h(t_n) - \widehat{Y}_h(t)\big]\,\mathrm{d}t - \int_{t_n}^{t_{n+1}} \widetilde{f}_h(t) - f_h(t)\,\mathrm{d}t\,.$$

Testing this equation with $(-\Delta_h)^{-1} e_n$, then applying discrete Gronwall's inequality, and finally taking the summation over n we arrive at

$$\begin{aligned} \max_{0\le k\le N} \|e_k\|^2_{\dot{\mathbb{H}}_h^{-1}} + \sum_{n=0}^{N-1} \tau \|e_n\|^2 \le C\Big[\|e_N\|^2_{\dot{\mathbb{H}}_h^{-1}} + \sum_{n=0}^{N-1} \int_{t_n}^{t_{n+1}} \|\widehat{Y}_h(t) - \widehat{Y}_h(t_n)\|^2\,\mathrm{d}t \\ + \int_0^T \|\widetilde{f}_h(t) - f_h(t)\|^2_{\dot{\mathbb{H}}_h^{-1}}\,\mathrm{d}t\Big]\,. \end{aligned} \tag{3.29}$$

To bound the second term on the right-hand side, we recall that $\widehat{Y}_h$ solves the *random* ODE (3.26). Hence we may expect order $O(\tau^2)$; in fact, by a simple argument as in Lemma 3.3, we get

$$\sum_{n=0}^{N-1} \int_{t_n}^{t_{n+1}} \|\widehat{Y}_h(t) - \widehat{Y}_h(t_n)\|^2\,\mathrm{d}t \le C\tau^2\Big[\|Y_{T,h}\|^2_{\dot{\mathbb{H}}_h^1} + \int_0^T \|f_h(t)\|^2\,\mathrm{d}t\Big]\,.$$

Consequently, we can conclude from (3.29) that

$$\begin{aligned} \max_{0\le n\le N} \|e_n\|^2_{\dot{\mathbb{H}}_h^{-1}} + \sum_{n=0}^{N-1} \tau \|e_n\|^2 &\le C\tau^2\Big[\|Y_{T,h}\|^2_{\dot{\mathbb{H}}_h^1} + \int_0^T \|f_h(t)\|^2\,\mathrm{d}t\Big] \\ &\quad + C\Big[\|e_N\|^2_{\dot{\mathbb{H}}_h^{-1}} + \int_0^T \|\widetilde{f}_h(t) - f_h(t)\|^2_{\dot{\mathbb{H}}_h^{-1}}\,\mathrm{d}t\Big]\,. \end{aligned} \tag{3.30}$$

By applying (3.27) and Jensen's inequality for conditional expectations, we have

$$\max_{0\le n\le N} \mathbb{E}\big[\|Y_n - Y_h(t_n)\|^2_{\dot{\mathbb{H}}_h^{-1}}\big] + \tau \sum_{n=0}^{N-1} \mathbb{E}\big[\|Y_n - Y_h(t_n)\|^2\big]$$

$$\le \mathbb{E}\Big[\max_{0\le k\le N} \|e_k\|^2_{\dot{\mathbb{H}}_h^{-1}} + \sum_{n=0}^{N-1} \tau \|e_n\|^2\Big],$$

which, together with (3.30), settles the assertion. □

We are now ready to verify rates for scheme (3.20) for general β.

Theorem 3.5 *Let* $\big(Y_h(\cdot), Z_h(\cdot)\big)$ *resp.* $\big(Y_\cdot, Z_0(\cdot)\big)$ *be solutions to* (3.13) *resp.* (3.20). *Then there exists a constant* C *such that*

$$\max_{0\le n\le N} \mathbb{E}\big[\|Y_h(t_n) - Y_n\|^2_{\dot{\mathbb{H}}_h^{-1}}\big] + \sum_{n=0}^{N-1} \mathbb{E}\Big[\int_{t_n}^{t_{n+1}} \|Y_h(t) - Y_n\|^2 + \|Z_h(t) - Z_0(t)\|^2_{\dot{\mathbb{H}}_h^{-1}} \,\mathrm{d}t\Big]$$

$$\le C\tau \mathbb{E}\Big[\|Y_{T,h}\|^2_{\dot{\mathbb{H}}_h^2} + \int_0^T \|f_h(t)\|^2_{\dot{\mathbb{H}}_h^1} \,\mathrm{d}t\Big] + C\mathbb{E}\Big[\|Y_{T,h} - \widetilde{Y}_{T,h}\|^2_{\dot{\mathbb{H}}_h^{-1}}$$

$$+ \sum_{n=0}^{N-1} \int_{t_n}^{t_{n+1}} \|Z_h(t) - \bar{Z}_h(t)\|^2_{\dot{\mathbb{H}}_h^{-1}} + \|f_h(t) - \widetilde{f}_h(t)\|^2_{\dot{\mathbb{H}}_h^{-1}} \,\mathrm{d}t\Big].$$

Proof We revisit Theorem 3.4 and view $\beta Z_h(\cdot) - f_h(\cdot)$ resp. $\beta \bar{Z}_0(\cdot) - \widetilde{f}_h(\cdot)$ as $f_h(\cdot)$ resp. $\widetilde{f}_h(\cdot)$. Together with estimate (3.15) in Lemma 3.3, the triangle inequality then leads to

$$\max_{0\le n\le N} \mathbb{E}\big[\|Y_n - Y_h(t_n)\|^2_{\dot{\mathbb{H}}_h^{-1}}\big] + \sum_{n=0}^{N-1} \mathbb{E}\Big[\int_{t_n}^{t_{n+1}} \|Y_h(t) - Y_n\|^2 \,\mathrm{d}t\Big]$$

$$\le \max_{0\le n\le N} \mathbb{E}\big[\|Y_n - Y_h(t_n)\|^2_{\dot{\mathbb{H}}_h^{-1}}\big]$$

$$+ C\sum_{n=0}^{N-1} \mathbb{E}\Big[\int_{t_n}^{t_{n+1}} \|Y_h(t) - Y_h(t_n)\|^2 + \|Y_h(t_n) - Y_n\|^2 \,\mathrm{d}t\Big]$$

$$\le C\tau \mathbb{E}\Big[\|Y_{T,h}\|^2_{\dot{\mathbb{H}}_h^1} + \int_0^T \|\beta Z_h(t) - f_h(t)\|^2 \,\mathrm{d}t\Big] + C\mathbb{E}\big[\|Y_{T,h} - \widetilde{Y}_{T,h}\|^2_{\dot{\mathbb{H}}_h^{-1}}\big]$$

$$+ C\sum_{n=0}^{N-1} \mathbb{E}\Big[\int_{t_n}^{t_{n+1}} \beta^2 \|\bar{Z}_0(t) - Z_h(t)\|^2_{\dot{\mathbb{H}}_h^{-1}} + \|f_h(t) - \widetilde{f}_h(t)\|^2_{\dot{\mathbb{H}}_h^{-1}} \,\mathrm{d}t\Big]. \quad (3.31)$$

In the next step, we estimate $\sum_{n=0}^{N-1} \mathbb{E}\big[\int_{t_n}^{t_{n+1}} \|\bar{Z}_0(t) - Z_h(t)\|^2_{\dot{\mathbb{H}}_h^{-1}} \,\mathrm{d}t\big]$. By the triangle inequality, the definition of $\bar{Z}_0(\cdot)$ in (3.21), and Hölder's inequality, it follows that

$$
\begin{aligned}
&\mathbb{E}\Big[\int_{t_n}^{t_{n+1}} \|\bar{Z}_0(t) - Z_h(t)\|^2_{\dot{\mathbb{H}}_h^{-1}} \, \mathrm{d}t\Big] \\
&\quad \leq C\,\mathbb{E}\Big[\int_{t_n}^{t_{n+1}} \|\bar{Z}_0(t) - \bar{Z}_h(t)\|^2_{\dot{\mathbb{H}}_h^{-1}} + \|Z_h(t) - \bar{Z}_h(t)\|^2_{\dot{\mathbb{H}}_h^{-1}} \, \mathrm{d}t\Big] \\
&\quad \leq C\,\mathbb{E}\Big[\int_{t_n}^{t_{n+1}} \|Z_0(t) - Z_h(t)\|^2_{\dot{\mathbb{H}}_h^{-1}} + \|Z_h(t) - \bar{Z}_h(t)\|^2_{\dot{\mathbb{H}}_h^{-1}} \, \mathrm{d}t\Big]. \qquad (3.32)
\end{aligned}
$$

Estimate (3.32) contains a term for $Z_0(\cdot) - Z_h(\cdot)$ that we need to bound; for this purpose, we go back to equations (3.13) and (3.20). Setting $e_n^Y = Y_n - Y_h(t_n)$ then leads to

$$
\begin{aligned}
&(\mathbb{1}_h - \tau\Delta_h)e_n^Y + \int_{t_n}^{t_{n+1}} Z_0(t) - Z_h(t)\,\mathrm{d}W(t) \\
&\quad = e_{n+1}^Y + \int_{t_n}^{t_{n+1}} \Delta_h\big[Y_h(t) - Y_h(t_n)\big] + \beta\big[\bar{Z}_0(t) - \bar{Z}_h(t)\big] \\
&\qquad + \beta\big[\bar{Z}_h(t) - Z_h(t)\big] + \big[f_h(t) - \widetilde{f}_h(t)\big]\,\mathrm{d}t\,.
\end{aligned}
$$

By the mutual independence of e_n^Y and $\int_{t_n}^{t_{n+1}} Z_0(t) - Z_h(t)\,\mathrm{d}W(t)$, upon squaring both sides, taking expectations and then applying Young's inequality as well as Hölder's inequality, we have ($\varepsilon > 0$)

$$
\begin{aligned}
&\mathbb{E}\big[\|(\mathbb{1}_h - \tau\Delta_h)e_n^Y\|^2_{\dot{\mathbb{H}}_h^{-1}}\big] + \mathbb{E}\Big[\int_{t_n}^{t_{n+1}} \|Z_0(t) - Z_h(t)\|^2_{\dot{\mathbb{H}}_h^{-1}} \, \mathrm{d}t\Big] \\
&\leq (1 + 4\varepsilon)\mathbb{E}\big[\|e_{n+1}^Y\|^2_{\dot{\mathbb{H}}_h^{-1}}\big] \\
&\quad + \Big[4 + \frac{1}{\varepsilon}\Big]\tau\mathbb{E}\Big[\int_{t_n}^{t_{n+1}} \big\|\Delta_h\big[Y_h(t) - Y_h(t_n)\big]\big\|^2_{\dot{\mathbb{H}}_h^{-1}} + \beta^2\|Z_0(t) - Z_h(t)\|^2_{\dot{\mathbb{H}}_h^{-1}} \\
&\qquad + \beta^2\|Z_h(t) - \bar{Z}_h(t)\|^2_{\dot{\mathbb{H}}_h^{-1}} + \|f_h(t) - \widetilde{f}_h(t)\|^2_{\dot{\mathbb{H}}_h^{-1}} \, \mathrm{d}t\Big].
\end{aligned}
$$

By choosing ε such that $\big[4 + \frac{1}{\varepsilon}\big]\beta^2\tau \leq \frac{1}{2}$ (for example, we may choose τ small enough satisfying $\tau \leq \frac{1}{16\beta^2}$, and take $\varepsilon = \frac{2\beta^2\tau}{1-8\beta^2\tau}$), we may absorb the third term on the right-hand side, such that

$$
\begin{aligned}
&\mathbb{E}\big[\|e_n^Y\|^2_{\dot{\mathbb{H}}_h^{-1}}\big] + \frac{1}{2}\mathbb{E}\Big[\int_{t_n}^{t_{n+1}} \|Z_0(t) - Z_h(t)\|^2_{\dot{\mathbb{H}}_h^{-1}} \, \mathrm{d}t\Big] \\
&\quad \leq \Big[1 + \frac{8\beta^2}{1 - 8\beta^2\tau}\tau\Big]\mathbb{E}\big[\|e_{n+1}^Y\|^2_{\dot{\mathbb{H}}_h^{-1}}\big] \\
&\qquad + \frac{1}{8\beta^2}\mathbb{E}\Big[\int_{t_n}^{t_{n+1}} \big\|\Delta_h\big[Y_h(t) - Y_h(t_n)\big]\big\|^2_{\dot{\mathbb{H}}_h^{-1}} + \beta^2\|Z_h(t) - \bar{Z}_h(t)\|^2_{\dot{\mathbb{H}}_h^{-1}} \\
&\qquad\quad + \|f_h(t) - \widetilde{f}_h(t)\|^2_{\dot{\mathbb{H}}_h^{-1}} \, \mathrm{d}t\Big].
\end{aligned}
$$

We now apply discrete Gronwall's inequality and taking the summation over n together with (3.15) for $\gamma = 1$ in Lemma 3.3 to conclude

$$\begin{aligned}
&\max_{k=0,1,\cdots,N} \mathbb{E}\big[\|e_k^Y\|^2_{\dot{\mathbb{H}}_h^{-1}}\big] + \mathbb{E}\Big[\int_0^T \|Z_0(t) - Z_h(t)\|^2_{\dot{\mathbb{H}}_h^{-1}}\,\mathrm{d}t\Big] \\
&\le C\tau\mathbb{E}\Big[\|Y_{T,h}\|^2_{\dot{\mathbb{H}}_h^2} + \int_0^T \|f_h(t)\|^2_{\dot{\mathbb{H}}_h^1}\,\mathrm{d}t\Big] + C\Big\{\mathbb{E}\big[\|Y_h(T) - Y_N\|^2_{\dot{\mathbb{H}}_h^{-1}}\big] \\
&+\sum_{n=0}^{N-1}\mathbb{E}\Big[\int_{t_n}^{t_{n+1}} \beta^2\|Z_h(t)-\bar{Z}_h(t)\|^2_{\dot{\mathbb{H}}_h^{-1}} + \|f_h(t) - \widetilde{f}_h(t)\|^2_{\dot{\mathbb{H}}_h^{-1}}\,\mathrm{d}t\Big]\Big\}. \qquad (3.33)
\end{aligned}$$

Finally, the desired assertion can be derived by combining with (3.31)–(3.33). □

Remark 3.3 Remark 3.2 motivates that auxiliary tuple $\big(Y_\cdot, Z_0(\cdot)\big)$ serves as a bridge between $\big(Y_h(\cdot), Z_h(\cdot)\big)$ and $\big(Y_{h\tau}(\cdot), Z_{h\tau}(\cdot)\big)$. By choosing $\widetilde{Y}_{T,h}$, $\widetilde{f}_h(\cdot)$ as in (3.22), and applying Theorem 3.5 and (3.32), we obtain the following error estimate for the Euler method (3.19):

$$\begin{aligned}
&\max_{0\le n\le N} \mathbb{E}\big[\|Y_h(t_n) - Y_{h\tau}(t_n)\|^2_{\dot{\mathbb{H}}_h^{-1}}\big] \\
&+\sum_{n=0}^{N-1}\mathbb{E}\Big[\int_{t_n}^{t_{n+1}} \|Y_h(t) - Y_{h\tau}(t_n)\|^2 + \|Z_h(t) - Z_{h\tau}(t_n)\|^2_{\dot{\mathbb{H}}_h^{-1}}\,\mathrm{d}t\Big] \\
&\le C\tau\mathbb{E}\Big[\|Y_{T,h}\|^2_{\dot{\mathbb{H}}_h^2} + \int_0^T \|f_h(t)\|^2_{\dot{\mathbb{H}}_h^1}\,\mathrm{d}t\Big] \\
&+C\mathbb{E}\Big[\sum_{n=0}^{N-1}\int_{t_n}^{t_{n+1}} \|Z_h(t) - \bar{Z}_h(t)\|^2_{\dot{\mathbb{H}}_h^{-1}} + \|f_h(t) - f_h(t_{n+1})\|^2_{\dot{\mathbb{H}}_h^{-1}}\,\mathrm{d}t\Big].
\end{aligned}$$

Thus, to get a rate, regularity in time for $Z_h(\cdot)$ is crucial. Generally speaking, there are two main cases to derive needed regularity for $Z_h(\cdot)$: 1) the concerned BSDEs are Markovian; 2) $Y_{T,h}$, $f_h(\cdot)$ have suitable Malliavin derivatives, i.e., $Y_{T,h}, f_h(t) \in \mathbb{D}^\gamma(\mathbb{H})$ for suitable γ , $\mathbb{H}$. See e.g. [7–9] and the references therein. In Sect. 3.3.3, we will use Malliavin calculus to derive a rate for the temporal discretization of BSDEs.

3.3 Discretization of Problem (SLQ) and Rates

We split the construction and convergence analysis of a numerical method to approximate Problem **(SLQ)** from Chap. 1 into two steps: in Sect. 3.3.1 we begin with Problem **(SLQ)**$_h$, which is a discretization scheme in space only; related stability bounds for its solution will be provided in Sect. 3.3.3 to verify rates of convergence for Problem **(SLQ)**$_{h\tau}$—the spatio-temporal discretization of Problem **(SLQ)**. Conceptually, we follow the approach to 'first discretize, then optimize'; see also [10].

The numerical analysis of Problem **(SLQ)** was also considered in [11–15]. Here, we proceed as follows to obtain optimal convergence rates:

(a) We use weaker assumptions on the data x and $\sigma(\cdot)$ as were imposed in [11, 14], and derive improved convergence rates in space.
(b) To construct an implementable algorithm for Problem **(SLQ)**, we adopt the Euler method for temporal discretization. Then by Malliavin calculus, we verify additional regularity of the optimal control $U_h^*(\cdot)$ of Problem $\textbf{(SLQ)}_h$ to prove an optimal convergence rate in time for it; see Sects. 3.3.2 and 3.3.3.

3.3.1 Spatial Discretization of Problem (SLQ)

In this part, we present a semi-discretized method in space of Problem **(SLQ)**, and prove a rate of convergence. Relying on the finite element method, the spatial discretization is stated as follows.

Problem $\textbf{(SLQ)}_h$. Find the optimal control $U_h^*(\cdot) \in L^2_{\mathbb{F}}(0, T; \mathbb{V}_h)$ that minimizes the cost functional

$$\mathcal{J}_h\big(U_h(\cdot)\big) = \frac{1}{2}\mathbb{E}\Big[\int_0^T \|X_h(t)\|^2 + \|U_h(t)\|^2\,\mathrm{d}t\Big] + \frac{\alpha}{2}\mathbb{E}\big[\|X_h(T)\|^2\big], \tag{3.34}$$

subject to the following SDE

$$\begin{cases} \mathrm{d}X_h(t) = \big[\Delta_h X_h(t) + U_h(t)\big]\,\mathrm{d}t + \big[\beta X_h(t) + \Pi_h\sigma(t)\big]\mathrm{d}W(t) & t \in (0, T]\,, \\ X_h(0) = \Pi_h x\,. \end{cases} \tag{3.35}$$

The existence of a unique optimal control $U_h^*(\cdot)$ follows from [16, Chap. 6]. Moreover, thanks to Pontryagin's maximum principle, $U_h^*(\cdot)$ enjoys an open-loop representation:

$$U_h^*(t) - Y_h(t) = 0 \qquad t \in [0, T]\,, \tag{3.36}$$

where $\big(Y_h(\cdot), Z_h(\cdot)\big) \in L^2_{\mathbb{F}}\big(\Omega; C([0, T]; \mathbb{V}_h)\big) \times L^2_{\mathbb{F}}\big(0, T; \mathbb{V}_h\big)$ solves the BSDE

$$\begin{cases} \mathrm{d}Y_h(t) = \big[-\Delta_h Y_h(t) - \beta Z_h(t) + X_h^*(t)\big]\,\mathrm{d}t + Z_h(t)\,\mathrm{d}W(t) & t \in [0, T)\,, \\ Y_h(T) = -\alpha X_h^*(T)\,. \end{cases} \tag{3.37}$$

The main result in this section is the following theorem which shows optimal rates of convergence for the optimal pair of Problem $\textbf{(SLQ)}_h$.

Theorem 3.6 *Suppose that* **(A)** *in part* **a.** *of Sect. 2.1 holds. Let* $\big(X^*(\cdot), U^*(\cdot)\big)$ *resp.* $\big(X_h^*(\cdot), U_h^*(\cdot)\big)$ *be the optimal pairs to Problems* **(SLQ)** *resp.* $\textbf{(SLQ)}_h$. *There exists* C *independent of* h *such that*

$$\mathbb{E}\Big[\int_0^T \|X^*(t) - X_h^*(t)\|^2 + \|U^*(t) - U_h^*(t)\|^2 \,\mathrm{d}t\Big] \tag{3.38}$$
$$\leq Ch^4\big[\|x\|_{\mathbb{H}_0^1}^2 + \|\sigma(\cdot)\|_{L_{\mathbb{F}}^2(0,T;\mathbb{H}_0^1)}^2\big],$$

and

$$\mathbb{E}\Big[\sup_{0\leq t\leq T} \|X^*(t) - X_h^*(t)\|^2\Big] \leq Ch^4\big[\|x\|_{\mathbb{H}_0^1\cap\mathbb{H}^2}^2 + \|\sigma(\cdot)\|_{L_{\mathbb{F}}^2(0,T;\mathbb{H}_0^1\cap\mathbb{H}^2)}^2\big]. \tag{3.39}$$

To prove Theorem 3.6, we introduce several operators $\mathcal{S}$, $\mathcal{T}^1$, $\mathcal{T}^2$ which are related to solution $\big(X(\cdot), Y(\cdot), Z(\cdot)\big)$ of FBSPDE (2.18). Specifically, the 'control-to-state' map $\mathcal{S} : L_{\mathbb{F}}^2(0, T; \mathbb{L}^2) \to L_{\mathbb{F}}^2\big(\Omega; C([0, T]; \mathbb{H}_0^1)\big) \cap L_{\mathbb{F}}^2(0, T; \mathbb{H}_0^1 \cap \mathbb{H}^2)$ is defined by $\mathcal{S}\big(U(\cdot)\big)(\cdot) = X\big(\cdot; x, U(\cdot)\big)$. If $x = 0$ and $\sigma(\cdot) = 0$, we denote this solution map by $\mathcal{S}^0$. We know that the solution to equation (2.18b) depends on $X^*(\cdot)$, which may be written as $\big(Y(\cdot; X^*(\cdot)), Z(\cdot; X^*(\cdot))\big) = \big(\mathcal{T}^1(X^*(\cdot))(\cdot), \mathcal{T}^2(X^*(\cdot))(\cdot)\big)$, where

$$\mathcal{T}^1 : L_{\mathbb{F}}^2\big(\Omega; C([0, T]; \mathbb{H}_0^1)\big) \to L_{\mathbb{F}}^2\big(\Omega; C([0, T]; \mathbb{H}_0^1)\big) \cap L_{\mathbb{F}}^2(0, T; \mathbb{H}_0^1 \cap \mathbb{H}^2),$$
$$\mathcal{T}^2 : L_{\mathbb{F}}^2\big(\Omega; C([0, T]; \mathbb{H}_0^1)\big) \to L_{\mathbb{F}}^2(0, T; \mathbb{H}_0^1).$$

To simplify notation, we shall use $\mathcal{S}(\cdot; U)$, $\mathcal{S}^0(\cdot; U)$, and $\mathcal{T}^i(\cdot; \mathcal{S}(U))$ in place of $\mathcal{S}(U(\cdot))(\cdot)$, $\mathcal{S}^0(U(\cdot))(\cdot)$, and $\mathcal{T}^i(\mathcal{S}(U(\cdot)))(\cdot)$, respectively, for $i = 1, 2$.

Lemma 3.5 *For every $U(\cdot) \in L_{\mathbb{F}}^2(0, T; \mathbb{L}^2)$, the Fréchet derivative $D\mathcal{J}\big(U(\cdot)\big)$ is a bounded operator on $L_{\mathbb{F}}^2(0, T; \mathbb{L}^2)$ which takes the form.*

$$D\mathcal{J}\big(U(\cdot)\big) = U(\cdot) - \mathcal{T}^1\big(\cdot; \mathcal{S}(U)\big). \tag{3.40}$$

Proof By Lemma 2.5 on the stability of SPDE (2.18a), we have

$$\mathbb{E}\big[\|\mathcal{S}^0(t; V)\|^2\big] \leq C\|V(\cdot)\|_{L_{\mathbb{F}}^2(0,T;\mathbb{L}^2)}^2 \qquad \forall t \in [0, T],\ V(\cdot) \in L_{\mathbb{F}}^2(0, T; \mathbb{L}^2).$$

Hence, for any $U(\cdot),\ V(\cdot) \in L_{\mathbb{F}}^2(0, T; \mathbb{L}^2)$, since the linearity of SPDE (2.18a) yields $\mathcal{S}(\cdot; U + V) = \mathcal{S}(\cdot; U) + \mathcal{S}^0(\cdot; V)$, we can get for the quadratic cost functional $\mathcal{J}(\cdot)$ defined in (1.1)

$$\mathcal{J}\big(U(\cdot) + V(\cdot)\big) - \mathcal{J}\big(U(\cdot)\big) - \Big[\big(\mathcal{S}(\cdot; U), \mathcal{S}^0(\cdot; V)\big)_{L_{\mathbb{F}}^2(0,T;\mathbb{L}^2)}$$
$$+\big(\alpha\mathcal{S}(T; U), \mathcal{S}^0(T; V)\big)_{L_{\mathcal{F}_T}^2(\Omega;\mathbb{L}^2)} + \big(U(\cdot), V(\cdot)\big)_{L_{\mathbb{F}}^2(0,T;\mathbb{L}^2)}\Big]$$
$$= \frac{1}{2}\Big[\big\|\mathcal{S}^0(\cdot; V)\big\|_{L_{\mathbb{F}}^2(0,T;\mathbb{L}^2)}^2 + \|V(\cdot)\|_{L_{\mathbb{F}}^2(0,T;\mathbb{L}^2)}^2 + \alpha\big\|\mathcal{S}^0(T; V)\big\|_{L_{\mathcal{F}_T}^2(\Omega;\mathbb{L}^2)}^2\Big].$$

On the other hand, Itô's formula to $\big(\mathcal{S}^0(t; V), \mathcal{T}^1\big(t; \mathcal{S}(U)\big)\big)_{\mathbb{L}^2}$ yields

$$\begin{aligned}&\big(\mathcal{S}(\cdot;U),\mathcal{S}^0(\cdot;V)\big)_{L^2_{\mathbb{F}}(0,T;\mathbb{L}^2)}+\big(\alpha\mathcal{S}(T;U),\mathcal{S}^0(T;V)\big)_{L^2_{\mathcal{F}_T}(\Omega;\mathbb{L}^2)}\\&\quad=-\big(\mathcal{T}^1\big(\cdot;\mathcal{S}(U)\big),V(\cdot)\big)_{L^2_{\mathbb{F}}(0,T;\mathbb{L}^2)}.\end{aligned}$$

By the definition of Fréchet derivative, the desired result now follows. □

Similar to the definition of $\mathcal{S}$, by the unique solvability of (3.35), we associate to this equation the bounded solution operator $\mathcal{S}_h : L^2_{\mathbb{F}}(0,T;\mathbb{V}_h) \to L^2_{\mathbb{F}}\big(\Omega; C([0,T];\mathbb{V}_h)\big)$. When $x = 0$ and $\sigma(\cdot) = 0$, we denote this solution operator by $\mathcal{S}^0_h$. Similarly, the solution pair to BSDE (3.37) may be written in the following form $\big(Y_h\big(\cdot;X^*_h(\cdot)\big), Z_h\big(\cdot;X^*_h(\cdot)\big)\big) = \big(\mathcal{T}^1_h\big(X^*_h(\cdot)\big)(\cdot), \mathcal{T}^2_h\big(X^*_h(\cdot)\big)(\cdot)\big)$, where

$$\begin{aligned}\mathcal{T}^1_h &: L^2_{\mathbb{F}}\big(\Omega; C([0,T];\mathbb{V}_h)\big) \to L^2_{\mathbb{F}}\big(\Omega; C([0,T];\mathbb{V}_h)\big),\\ \mathcal{T}^2_h &: L^2_{\mathbb{F}}\big(\Omega; C([0,T];\mathbb{V}_h)\big) \to L^2_{\mathbb{F}}(0,T;\mathbb{V}_h).\end{aligned}$$

Like the simplified notations $\mathcal{S}(\cdot;U), \mathcal{S}^0(\cdot;U), \mathcal{T}^i\big(\cdot;\mathcal{S}(U)\big)$ for $i = 1,2$, we here also introduce $\mathcal{S}_h(\cdot;U_h), \mathcal{S}^0_h(\cdot;U_h), \mathcal{T}^i_h\big(\cdot;\mathcal{S}_h(U_h)\big)$.

We are now in a position to prove Theorem 3.6.

Proof of Theorem 3.6.

(1) Verification of (3.38). For any control variable $U_h(\cdot) \in L^2_{\mathbb{F}}(0,T;\mathbb{V}_h)$, the Fréchet derivative $D\mathcal{J}_h\big(U_h(\cdot)\big)$ is a bounded operator (uniformly in h) on $L^2_{\mathbb{F}}(0,T;\mathbb{V}_h)$, and has the form

$$D\mathcal{J}_h\big(U_h(\cdot)\big) = U_h(\cdot) - \mathcal{T}^1_h\big(\cdot;\mathcal{S}_h(U_h)\big), \tag{3.41}$$

which can be deduced by a similar procedure as in the proof of Lemma 3.5. Let $U_h(\cdot), R_h(\cdot) \in L^2_{\mathbb{F}}(0,T;\mathbb{V}_h)$ be arbitrary; it is due to the quadratic structure of the cost functional $\mathcal{J}_h(\cdot)$ given in (3.34) that

$$D^2\mathcal{J}_h\big(U_h(\cdot)\big)R_h(\cdot) = R_h(\cdot) - \mathcal{T}^1_h\big(\cdot;\mathcal{S}^0_h(R_h)\big).$$

Note that the right-hand side of the above equation is independent of $U_h(\cdot)$. Then by Itô's formula to $\big(\mathcal{S}^0_h(t;R_h), \mathcal{T}^1_h\big(t;\mathcal{S}^0_h(R_h)\big)\big)_{\mathbb{L}^2}$ we get

$$\begin{aligned}&\Big(\mathcal{T}^1_h\big(\cdot;\mathcal{S}^0_h(R_h)\big), R_h(\cdot)\Big)_{L^2_{\mathbb{F}}(0,T;\mathbb{L}^2)}\\&= -\alpha\big(\mathcal{S}^0_h(T;R_h), \mathcal{S}^0_h(T;R_h)\big)_{L^2_{\mathcal{F}_T}(\Omega;\mathbb{L}^2)} - \big(\mathcal{S}^0_h(\cdot;R_h), \mathcal{S}^0_h(\cdot;R_h)\big)_{L^2_{\mathbb{F}}(0,T;\mathbb{L}^2)},\end{aligned}$$

and therefore

$$
\begin{aligned}
&\Big(D^2\mathcal{J}_h\big(U_h(\cdot)\big)R_h(\cdot),\, R_h(\cdot)\Big)_{L^2_{\mathbb{F}}(0,T;\mathbb{L}^2)} \\
&= \|R_h(\cdot)\|^2_{L^2_{\mathbb{F}}(0,T;\mathbb{L}^2)} + \|\mathcal{S}^0_h(\cdot;\, R_h)\|^2_{L^2_{\mathbb{F}}(0,T;\mathbb{L}^2)} + \alpha\|\mathcal{S}^0_h(T;\, R_h)\|^2_{L^2_{\mathcal{F}_T}(\Omega;\mathbb{L}^2)} \\
&\geq \|R_h(\cdot)\|^2_{L^2_{\mathbb{F}}(0,T;\mathbb{L}^2)} \qquad \forall\, R_h(\cdot) \in L^2_{\mathbb{F}}(0,T;\, \mathbb{V}_h)\,.
\end{aligned}
$$

As a consequence, on putting $R_h(\cdot) = U^*_h(\cdot) - \Pi_h U^*(\cdot)$, we have

$$
\begin{aligned}
\|U^*_h(\cdot) - \Pi_h U^*(\cdot)\|^2_{L^2_{\mathbb{F}}(0,T;\mathbb{L}^2)} \leq\, & \big(D\mathcal{J}_h\big(U^*_h(\cdot)\big),\, U^*_h(\cdot) - \Pi_h U^*(\cdot)\big)_{L^2_{\mathbb{F}}(0,T;\mathbb{L}^2)} \\
& - \big(D\mathcal{J}_h\big(\Pi_h U^*(\cdot)\big),\, U^*_h(\cdot) - \Pi_h U^*(\cdot)\big)_{L^2_{\mathbb{F}}(0,T;\mathbb{L}^2)}\,.
\end{aligned}
$$

and then

$$
\|U^*_h(\cdot) - \Pi_h U^*(\cdot)\|_{L^2_{\mathbb{F}}(0,T;\mathbb{L}^2)} \leq \big\|D\mathcal{J}_h\big(U^*_h(\cdot)\big) - D\mathcal{J}_h\big(\Pi_h U^*(\cdot)\big)\big\|_{L^2_{\mathbb{F}}(0,T;\mathbb{L}^2)}\,.
$$

Noticing that $D\mathcal{J}_h\big(U^*_h(\cdot)\big) = 0$ by (3.36), as well as $D\mathcal{J}\big(U^*(\cdot)\big) = 0$ by (2.19), we find that

$$
\begin{aligned}
&\|U^*_h(\cdot) - \Pi_h U^*(\cdot)\|^2_{L^2_{\mathbb{F}}(0,T;\mathbb{L}^2)} \\
&\leq 2\Big[\big\|D\mathcal{J}\big(U^*(\cdot)\big) - D\mathcal{J}\big(\Pi_h U^*(\cdot)\big)\big\|^2_{L^2_{\mathbb{F}}(0,T;\mathbb{L}^2)} \\
&\quad + \big\|D\mathcal{J}\big(\Pi_h U^*(\cdot)\big) - D\mathcal{J}_h\big(\Pi_h U^*(\cdot)\big)\big\|^2_{L^2_{\mathbb{F}}(0,T;\mathbb{L}^2)}\Big] \\
&=: 2(I_1 + I_2)\,.
\end{aligned}
\tag{3.42}
$$

In the following, we bound these two terms independently.

We use (3.40) to bound I_1 as follows,

$$
\begin{aligned}
I_1 \leq 2\Big[&\|U^*(\cdot) - \Pi_h U^*(\cdot)\|^2_{L^2_{\mathbb{F}}(0,T;\mathbb{L}^2)} \\
&+ \big\|\mathcal{T}^1\big(\cdot;\, \mathcal{S}(\Pi_h U^*)\big) - \mathcal{T}^1\big(\cdot;\, \mathcal{S}(U^*)\big)\big\|^2_{L^2_{\mathbb{F}}(0,T;\mathbb{L}^2)}\Big].
\end{aligned}
$$

By stability property (2.17) in Lemma 2.6 for BSPDE (2.18b) with data $Y(T) = -\alpha\big[\mathcal{S}(T;\, \Pi_h U^*) - \mathcal{S}(T;\, U^*)\big]$, $f(\cdot) = \mathcal{S}(\cdot;\, \Pi_h U^*) - \mathcal{S}(\cdot;\, U^*)$, as well as (2.13) in Lemma 2.5 for SPDE (2.18a) with $x = 0$, $f(\cdot) = U^*(\cdot) - \Pi_h U^*(\cdot)$, $g(\cdot) = 0$, the second term in the above inequality may be bounded by

$$
\begin{aligned}
&\leq C\Big[\|\mathcal{S}(T;\, U^*) - \mathcal{S}(T;\, \Pi_h U^*)\|^2_{L^2_{\mathcal{F}_T}(\Omega;\dot{\mathbb{H}}^{-1})} \\
&\quad + \|\mathcal{S}(\cdot;\, U^*) - \mathcal{S}(\cdot;\, \Pi_h U^*)\|^2_{L^2_{\mathbb{F}}(0,T;\dot{\mathbb{H}}^{-2})}\Big] \\
&\leq C\|U^*(\cdot) - \Pi_h U^*(\cdot)\|^2_{L^2_{\mathbb{F}}(0,T;\dot{\mathbb{H}}^{-1})}\,.
\end{aligned}
\tag{3.43}
$$

By optimality condition (2.19), and the regularity properties of the solution of FBSPDE (2.18), we know that $U^*(\cdot) = Y(\cdot) \in L^2_{\mathbb{F}}(0, T; \mathbb{H}^1_0 \cap \mathbb{H}^2)$; as a consequence, combining Lemma 2.2 with (2.23b) in Lemma 2.7 leads to

$$\begin{aligned} I_1 &\le Ch^4 \|Y(\cdot)\|^2_{L^2_{\mathbb{F}}(0,T;\mathbb{H}^1_0\cap\mathbb{H}^2)} \\ &\le Ch^4\big[\|X^*(T)\|^2_{L^2_{\mathcal{F}_T}(\Omega;\mathbb{H}^1_0)} + \|X^*(\cdot)\|^2_{L^2_{\mathbb{F}}(0,T;\mathbb{L}^2)}\big] \\ &\le Ch^4\big[\|x\|^2_{\mathbb{H}^1_0} + \|\sigma(\cdot)\|^2_{L^2(0,T;\mathbb{H}^1_0)}\big] . \end{aligned}$$

In the below, we apply representations (3.40), (3.41) to bound I_2 via

$$\begin{aligned} I_2 \le 2\Big[&\big\|\mathcal{T}^1(\cdot; \mathcal{S}(\Pi_h U^*)) - \mathcal{T}^1(\cdot; \mathcal{S}_h(\Pi_h U^*))\big\|^2_{L^2_{\mathbb{F}}(0,T;\mathbb{L}^2)} \\ &+\big\|\mathcal{T}^1(\cdot; \mathcal{S}_h(\Pi_h U^*)) - \mathcal{T}^1_h(\cdot; \mathcal{S}_h(\Pi_h U^*))\big\|^2_{L^2_{\mathbb{F}}(0,T;\mathbb{L}^2)}\Big] \\ =:&\, 2(I_{21} + I_{22}) . \end{aligned}$$

In order to bound I_{21}, we use the stability properties in Lemma 2.6 for BSPDE (2.18b), in combination with Theorem 3.1 on the error estimate for SPDE (2.18a) (note that $f(\cdot) = f_h(\cdot) = \Pi_h U^*(\cdot), g(\cdot) = \sigma(\cdot) , g_h(\cdot) = \Pi_h\sigma(\cdot)$), as well as (2.23b) in Lemma 2.7 to conclude that

$$\begin{aligned} I_{21} &\le C\Big[\big\|\mathcal{S}(\cdot; \Pi_h U^*) - \mathcal{S}_h(\cdot; \Pi_h U^*)\big\|^2_{L^2_{\mathbb{F}}(0,T;\dot{\mathbb{H}}^{-1})} \\ &\qquad +\big\|\mathcal{S}(T; \Pi_h U^*) - \mathcal{S}_h(T; \Pi_h U^*)\big\|^2_{L^2_{\mathcal{F}_T}(\Omega;\dot{\mathbb{H}}^{-2})}\Big] \\ &\le Ch^4\big[\|x\|^2_{\mathbb{H}^1_0} + \|\sigma(\cdot)\|^2_{L^2(0,T;\mathbb{H}^1_0)}\big] . \end{aligned}$$

For I_{22}, we use Theorem 3.3 with $Y_T = Y_{T,h} = -\alpha\mathcal{S}_h(T; \Pi_h U^*) , f(\cdot) = f_h(\cdot) = \mathcal{S}_h(\cdot; \Pi_h U^*)$ on the error estimate for BSPDE (2.18b), and then Lemma 3.1 with $f_h(\cdot) = \Pi_h U^*(\cdot)$, $g_h(\cdot) = \Pi_h\sigma(\cdot)$ on the stability properties for SDE (3.35) to find

$$\begin{aligned} I_{22} &\le Ch^4\big[\alpha^2\|\mathcal{S}_h(T; \Pi_h U^*)\|^2_{L^2_{\mathcal{F}_T}(\Omega;\mathbb{H}^1_0)} + \|\mathcal{S}_h(\cdot; \Pi_h U^*)\|^2_{L^2_{\mathbb{F}}(0,T;\mathbb{L}^2)}\big] . \\ &\le Ch^4\big[\|x\|^2_{\mathbb{H}^1_0} + \|\sigma(\cdot)\|^2_{L^2(0,T;\mathbb{H}^1_0)}\big] . \end{aligned}$$

We now insert these estimates into (3.42), and utilize the optimality condition (2.19) to obtain the bound

$$\begin{aligned} &\|U^*(\cdot) - U^*_h(\cdot)\|^2_{L^2_{\mathbb{F}}(0,T;\mathbb{L}^2)} \\ &\le 2\Big[\|U^*(\cdot) - \Pi_h U^*(\cdot)\|^2_{L^2_{\mathbb{F}}(0,T;\mathbb{L}^2)} + \|U^*_h(\cdot) - \Pi_h U^*(\cdot)\|^2_{L^2_{\mathbb{F}}(0,T;\mathbb{L}^2)}\Big] \\ &\le Ch^4\Big[\|Y(\cdot)\|^2_{L^2_{\mathbb{F}}(0,T;\mathbb{H}^1_0\cap\mathbb{H}^2)} + \|x\|^2_{\mathbb{H}^1_0} + \|\sigma(\cdot)\|^2_{L^2(0,T;\mathbb{H}^1_0)}\Big] \\ &\le Ch^4\Big[\|x\|^2_{\mathbb{H}^1_0} + \|\sigma(\cdot)\|^2_{L^2(0,T;\mathbb{H}^1_0)}\Big] . \end{aligned}$$

This is the estimate for $U^*(\cdot) - U_h^*(\cdot)$ in assertion (3.38) in Theorem 3.6. Then applying Theorem 3.1 by setting $f(\cdot) = U^*(\cdot)$, $f_h(\cdot) = U_h^*(\cdot)$, $g(\cdot) = \sigma(\cdot)$, $g_h(\cdot) = \Pi_h\sigma(\cdot)$, we have

$$\|X^*(\cdot) - X_h^*(\cdot)\|^2_{L^2_{\mathbb{F}}(0,T;\mathbb{L}^2)} \le Ch^4\big[\|x\|^2_{\mathbb{H}^1_0} + \|\sigma(\cdot)\|^2_{L^2(0,T;\mathbb{H}^1_0)}\big],$$

which is the estimate for $X^*(\cdot) - X_h^*(\cdot)$ in assertion (3.38).

(2) Verification of (3.39). By applying Theorem 3.1 with the choices $f(\cdot) = U^*(\cdot)$, $f_h(\cdot) = U_h^*(\cdot)$, $g(\cdot) = \sigma(\cdot)$, and $g_h(\cdot) = \Pi_h\sigma(\cdot)$, togethor with the assertion (3.38), we obtain

$$\begin{aligned}
&\mathbb{E}\Big[\sup_{t\in[0,T]} \|X^*(t) - X_h^*(t)\|^2\Big] \\
&\le Ch^4\Big\{\|x\|^2_{\mathbb{H}^1_0\cap\mathbb{H}^2} + \mathbb{E}\Big[\int_0^T \|U^*(t)\|^2_{\mathbb{H}^1_0\cap\mathbb{H}^2} + \|\sigma(t)\|^2_{\mathbb{H}^1_0\cap\mathbb{H}^2}\,\mathrm{d}t\Big]\Big\} \\
&\quad + C\,\mathbb{E}\Big[\int_0^T \|U^*(t) - U_h^*(t)\|^2\,\mathrm{d}t\Big] \\
&\le Ch^4\big[\|x\|^2_{\mathbb{H}^1_0\cap\mathbb{H}^2} + \|\sigma(\cdot)\|^2_{L^2(0,T;\mathbb{H}^1_0\cap\mathbb{H}^2)}\big].
\end{aligned}$$

That completes the proof of assertion (3.39).

Remark 3.4 We may use Theorem 3.6 to verify further error bounds,

$$\begin{aligned}
&\mathbb{E}\Big[\sup_{t\in[0,T]} \|X^*(t) - X_h^*(t)\|^2\Big] + \mathbb{E}\Big[\int_0^T \|X^*(t) - X_h^*(t)\|^2_{\mathbb{H}^1_0}\,\mathrm{d}t\Big] \\
&\qquad\qquad \le Ch^2\big[\|x\|^2_{\mathbb{H}^1_0} + \|\sigma(\cdot)\|^2_{L^2(0,T;\mathbb{H}^1_0)}\big], \qquad (3.44)\\
&\mathbb{E}\Big[\sup_{t\in[0,T]} \|Y(t)-Y_h(t)\|^2\Big] + \mathbb{E}\Big[\int_0^T \|Y(t)-Y_h(t)\|^2_{\mathbb{H}^1_0} + \|Z(t)-Z_h(t)\|^2\,\mathrm{d}t\Big] \\
&\qquad\qquad \le Ch^2\big[\|x\|^2_{\mathbb{H}^1_0} + \|\sigma(\cdot)\|^2_{L^2(0,T;\mathbb{H}^1_0)}\big], \qquad (3.45)
\end{aligned}$$

and

$$\sup_{t\in[0,T]} \mathbb{E}\big[\|Y(t) - Y_h(t)\|^2\big] \le Ch^4\big[\|x\|^2_{\mathbb{H}^1_0\cap\mathbb{H}^2} + \|\sigma(\cdot)\|^2_{L^2(0,T;\mathbb{H}^1_0\cap\mathbb{H}^2)}\big]. \qquad (3.46)$$

Indeed, Theorem 3.1 with $f(\cdot) = U^*(\cdot)$, $f_h(\cdot) = U_h^*(\cdot)$, $g(\cdot) = \sigma(\cdot)$, $g_h(\cdot) = \Pi_h\sigma(\cdot)$, and assertion (3.38) easily settle (3.44).

Applying (3.16) in Theorem 3.3 with $Y(T) = -\alpha X^*(T)$, $Y_h(T) = -\alpha X_h^*(T)$, and $f(\cdot) = X^*(\cdot)$, $f_h(\cdot) = X_h^*(\cdot)$, together with (2.4c) in Lemma 2.2 and assertion (3.38), yields (3.45).

Similarly, by (3.17) in Theorem 3.3, Lemma 2.5 with $f(\cdot) = U^*(\cdot)$, $g(\cdot) = \sigma(\cdot)$, $\gamma = 2$, (2.23b) in Lemma 2.7, and assertion (3.39), we arrive at (3.46).

3.3.2 Bounds in Strong Norms for the Optimal Pair to Problem $(\mathbf{SLQ})_h$

In this part, we verify uniform (in h) bounds in stronger norms for the optimal pair $\big(X_h^*(\cdot), U_h^*(\cdot)\big)$ to Problem $(\mathbf{SLQ})_h$, which will be applied in the error analysis of the temporal discretization of Problem $(\mathbf{SLQ})_h$; see Remark 3.3.

To begin with, we introduce a family of SLQ problems, which are parametrized by $t \in [0, T]$; for this purpose, we consider the controlled SDE

$$\begin{cases} \mathrm{d}X_h(s) = \big[\Delta_h X_h(s) + U_h(s)\big]\,\mathrm{d}s + \big[\beta X_h(s) + \Pi_h \sigma(s)\big]\,\mathrm{d}W(s) & s \in (t, T], \\ X_h(t) = \Pi_h x \end{cases} \tag{3.47}$$

with $x \in \mathbb{H}_0^1 \cap \mathbb{H}^2$, and the (parametrized) cost functional

$$\mathcal{J}_h\big(t, x; U_h(\cdot)\big) = \frac{1}{2}\mathbb{E}\Big[\int_t^T \|X_h(s)\|^2 + \|U_h(s)\|^2\,\mathrm{d}s\Big] + \frac{\alpha}{2}\mathbb{E}\big[\|X_h(T)\|^2\big]. \tag{3.48}$$

Now we define auxiliary SLQ problems as follows.

Problem $(\mathbf{SLQ})_{\text{aux}}^{t;h}$. For any given $t \in [0, T)$ and $x \in \mathbb{L}^2$, search for an optimal control $U_h^*(\cdot) \in L_{\mathbb{F}}^2(t, T; \mathbb{V}_h)$ such that

$$\mathcal{J}_h\big(t, x; U_h^*(\cdot)\big) = \inf_{U_h(\cdot) \in L_{\mathbb{F}}^2(t, T; \mathbb{V}_h)} \mathcal{J}_h\big(t, x; U_h(\cdot)\big). \tag{3.49}$$

It is clear that $\mathcal{J}_h\big(U_h(\cdot)\big) = \mathcal{J}_h\big(0, x; U_h(\cdot)\big)$. The unique solvability of Problem $(\mathbf{SLQ})_{\text{aux}}^{t;h}$ is shown in [16]. Moreover, the stochastic Riccati equation related to Problem $(\mathbf{SLQ})_{\text{aux}}^{t;h}$ reads:

$$\begin{cases} \mathcal{P}_h'(t) + \mathcal{P}_h(t)\Delta_h + \Delta_h \mathcal{P}_h(t) + \beta^2 \mathcal{P}_h(t) + \mathbb{1}_h - \mathcal{P}_h^2(t) = 0 & t \in [0, T], \\ \mathcal{P}_h(T) = \alpha \mathbb{1}_h, \end{cases} \tag{3.50}$$

and we consider a backward ODE

$$\begin{cases} \eta_h'(t) + \big[\Delta_h - \mathcal{P}_h(t)\big]\eta_h(t) + \mathcal{P}_h(t)\Pi_h \sigma(t) = 0 & t \in [0, T], \\ \eta_h(T) = 0. \end{cases} \tag{3.51}$$

By [16, Chap. 6, Theorems 6.1, 7.2], the stochastic Riccati equation (3.50) admits a unique solution $\mathcal{P}_h(\cdot) \in C\big([0, T]; \mathbb{S}_+(\mathbb{V}_h)\big)$, subsequently (3.51) has a unique solution $\eta_h(\cdot) \in C([0, T]; \mathbb{V}_h)$, and that

$$
\begin{aligned}
\mathcal{J}_h\big(t,x;U_h^*(\cdot)\big) =& \frac{1}{2}\big(\mathcal{P}_h(t)\Pi_h x,\,\Pi_h x\big)_{\mathbb{L}^2} + \big(\eta_h(t),\,\Pi_h x\big)_{\mathbb{L}^2} \\
&+ \frac{1}{2}\int_t^T \Big[\big(\mathcal{P}_h(s)\Pi_h\sigma(s),\,\Pi_h\sigma(s)\big)_{\mathbb{L}^2} + \|\eta_h(s)\|^2\Big]\,\mathrm{d}s\,,
\end{aligned}
\tag{3.52}
$$

where $U_h^*(\cdot)$ is the optimal control owning the following state feedback form

$$
U_h^*(t) = -\mathcal{P}_h(t)X_h^*(t) - \eta_h(t) \qquad t\in[0,T]\,. \tag{3.53}
$$

Lemma 3.6 *Let $\mathcal{P}_h(\cdot)$ be the solution to Riccati equation* (3.50), *and $\eta_h(\cdot)$ solves* (3.51). *There exists a constant C independent of h such that*

$$
\sup_{t\in[0,T]} \|\mathcal{P}_h(t)\|_{\mathcal{L}(\mathbb{L}^2|_{\mathbb{V}_h})} \le C\,, \tag{3.54}
$$

and

$$
\begin{aligned}
&\sup_{t\in[0,T]} \|\eta_h(t)\|^2 + \int_0^T \Big[\|\nabla\eta_h(t)\|^2 + \big(\mathcal{P}_h(t)\eta_h(t),\,\eta_h(t)\big)_{\mathbb{L}^2}\Big]\,\mathrm{d}t \\
&\qquad \le C\|\sigma(\cdot)\|^2_{L^2(0,T;\mathbb{L}^2)}\,.
\end{aligned}
\tag{3.55}
$$

Proof (1) Verification of (3.54). We consider Problem $\textbf{(SLQ)}^{t;h}_{\text{aux}}$ for (3.47) with $\sigma(\cdot)=0$. In this case, it follows that $\eta_h(\cdot)=0$. For the control variable $U_h(\cdot)=0$, Lemma 3.1 yields

$$
\sup_{s\in[t,T]} \mathbb{E}\big[\|X_h(s;\,\Pi_h x,0)\|^2\big] \le C\,\mathbb{E}\big[\|\Pi_h x\|^2\big].
$$

Hence, (3.49) and (3.52) lead to

$$
\big(\mathcal{P}_h(t)\Pi_h x,\,\Pi_h x\big)_{\mathbb{L}^2} \le 2\mathcal{J}_h(t,x;0) \le C\,\mathbb{E}\big[\|\Pi_h x\|^2\big]\,,
$$

which, together with the facts that $\mathcal{P}_h(\cdot)$ is nonnegative and Π_h is surjective, implies assertion (3.54).

(2) Verification of (3.55). Since $\mathcal{P}_h(\cdot)$ is nonnegative, we infer from (3.51) that

$$
\begin{aligned}
&\|\eta_h(t)\|^2 + 2\int_t^T \Big[\|\nabla\eta_h(s)\|^2 + \big(\mathcal{P}_h(s)\eta_h(s),\,\eta_h(s)\big)_{\mathbb{L}^2}\Big]\,\mathrm{d}s \\
&\quad \le \int_t^T \|\eta_h(s)\|^2\,\mathrm{d}s + \int_t^T \|\mathcal{P}_h(s)\Pi_h\sigma(s)\|^2\,\mathrm{d}s\,.
\end{aligned}
$$

Then Gronwall's inequality and (3.54) settle the assertion. □

The following lemma collects bounds for the solution $X_h^*(\cdot)$ and its Malliavin derivatives $D_\theta X_h^*(\cdot)$, $D_\mu D_\theta X_h^*(\cdot)$ to SDE (3.47) with $U_h(\cdot)=U_h^*(\cdot)$, by exploiting the state feedback representation (3.53) of the optimal control $U_h^*(\cdot)$, and the bounds for the stochastic Riccati operator in Lemma 3.6, in particular. These bounds will be applied to derive a convergence rate for an open-loop based scheme.

Lemma 3.7 *Let $\big(X_h^*(\cdot), U_h^*(\cdot)\big)$ be the optimal pair of Problem* **(SLQ)**$_h$. *Then it holds that $X_h^*(t) \in \mathbb{D}^{2,2}(\mathbb{L}^2)$ for any $t \in [0, T]$, and there exists a constant C independent of h such that*

$$\begin{cases} \mathbb{E}\Big[\sup_{t\in[0,T]} \|X_h^*(t)\|^2_{\dot{\mathbb{H}}_h^\gamma}\Big] + \mathbb{E}\Big[\int_0^T \|X_h^*(t)\|^2_{\dot{\mathbb{H}}_h^{\gamma+1}}\,dt\Big] \\ \qquad \le C\big[\|x\|^2_{\dot{\mathbb{H}}^\gamma} + \|\sigma(\cdot)\|^2_{L^2(0,T;\dot{\mathbb{H}}^\gamma)}\big] \quad \gamma = 0, 1\,, & (3.56a) \\ \mathbb{E}\Big[\sup_{t\in[0,T]} \|U_h^*(t)\|^2\Big] + \mathbb{E}\Big[\int_0^T \|U_h^*(t)\|^2_{\dot{\mathbb{H}}_h^1}\,dt\Big] \\ \qquad \le C\big[\|x\|^2 + \|\sigma(\cdot)\|^2_{L^2(0,T;\mathbb{L}^2)}\big]\,, & (3.56b) \\ \Big(\sup_{t\in[0,T]} \mathbb{E}\big[\|U_h^*(t)\|^4\big]\Big)^{1/2} \le C\big[\|x\|^2 + \|\sigma(\cdot)\|^2_{C([0,T];\mathbb{L}^2)}\big]\,, & (3.56c) \end{cases}$$

$$\begin{cases} \sup_{\theta\in[0,T]} \mathbb{E}\Big[\sup_{t\in[\theta,T]} \|D_\theta X_h^*(t)\|^2\Big] + \sup_{\theta,\mu\in[0,T]} \mathbb{E}\Big[\sup_{t\in[\mu\vee\theta,T]} \|D_\mu D_\theta X_h^*(t)\|^2\Big] \\ \qquad \le C\big[\|x\|^2 + \|\sigma(\cdot)\|^2_{C([0,T];\mathbb{L}^2)}\big]\,, & (3.57a) \\ \sup_{\theta\in[0,T]} \Big\{\mathbb{E}\Big[\sup_{t\in[\theta,T]} \|D_\theta X_h^*(t)\|^2_{\dot{\mathbb{H}}_h^1}\Big] + \mathbb{E}\Big[\int_\theta^T \|D_\theta X_h^*(t)\|^2_{\dot{\mathbb{H}}_h^2}\,dt\Big]\Big\} \\ \qquad \le C\big[\|x\|^2_{\mathbb{H}_0^1} + \|\sigma(\cdot)\|^2_{C([0,T];\mathbb{H}_0^1)}\big]\,, & (3.57b) \end{cases}$$

and

$$\begin{cases} \mathbb{E}\big[\|X_h^*(t) - X_h^*(s)\|^2\big] \\ \qquad \le C|t-s|\big[\|x\|^2_{\mathbb{H}_0^1} + \|\sigma(\cdot)\|^2_{C([0,T];\mathbb{L}^2)\cap L^2(0,T;\mathbb{H}_0^1)}\big] \quad t, s \in [0, T]\,, & (3.58a) \\ \mathbb{E}\Big[\sup_{t\in[\theta_1,T]} \|(D_{\theta_1} - D_{\theta_2})X_h^*(t)\|^2\Big] \\ \qquad \le C(\theta_1-\theta_2)\big[\|x\|^2_{\mathbb{H}_0^1} + \|\sigma(\cdot)\|^2_{C([0,T];\mathbb{H}_0^1)} + L^2_{\sigma,1/2}\big] \quad \theta_2 \le \theta_1\,. & (3.58b) \end{cases}$$

Proof (1) Verification of (3.56a)–(3.56c). We may follow the same steps that lead to (2.26), (2.23c), (2.23b) and (2.23a).

(2) Verification of (3.58a). Based on (3.56a)–(3.56b), the Itô isometry together with the triangle inequality implies that for $s \le t$,

$$\begin{aligned}\mathbb{E}\big[\|X_h^*(t) - X_h^*(s)\|^2\big] &\le C\Big\{|t-s|\mathbb{E}\Big[\int_0^T \|\Delta_h X_h^*(\theta)\|^2 + \|U_h^*(\theta)\|^2\,\mathrm{d}\theta\Big] \\ &\qquad + \mathbb{E}\Big[\int_s^t \|\beta X_h^*(\theta)\|^2 + \|\Pi_h\sigma(\theta)\|^2\,\mathrm{d}\theta\Big]\Big\} \\ &\le C|t-s|\big[\|x\|^2_{\mathbb{H}_0^1} + \|\sigma(\cdot)\|^2_{L^2(0,T;\mathbb{H}_0^1)\cap C([0,T];\mathbb{L}^2)}\big]\,.\end{aligned}$$

(3) Verification of (3.57a). To estimate the first term on the left-hand side, we use the facts that $\mathcal{P}_h(\cdot)$ and $\eta_h(\cdot)$ are deterministic, and [17, Theorem 2.2.1] to conclude that the first Malliavin derivative of $X_h^*(\cdot)$ exists. We take the Malliavin derivative on both sides of (3.47), together with the sate feedback control (3.53) to get

$$\begin{cases} \mathrm{d}D_\theta X_h^*(t) = \big[\Delta_h - \mathcal{P}_h(t)\big]D_\theta X_h^*(t)\,\mathrm{d}t + \beta D_\theta X_h^*(t)\,\mathrm{d}W(t) & t \in (\theta, T]\,, \\ D_\theta X_h^*(\theta) = \beta X_h^*(\theta) + \Pi_h\sigma(\theta)\,, \\ D_\theta X_h^*(t) = 0 \qquad t \in [0,\theta)\,. \end{cases} \tag{3.59}$$

Lemma 3.1 and the fact that $\mathcal{P}_h(\cdot) \in C\big([0,T]; \mathbb{S}_+(\mathbb{V}_h)\big)$ lead to

$$\begin{aligned}\mathbb{E}\Big[\sup_{t\in[\theta,T]} \|D_\theta X_h^*(t)\|^2\Big] &\le C\,\mathbb{E}\big[\|X_h^*(\theta) + \Pi_h\sigma(\theta)\|^2\big] \\ &\le C\big[\|x\|^2 + \|\sigma(\cdot)\|^2_{C([0,T];\mathbb{L}^2)}\big]\,,\end{aligned}$$

where C is independent of θ. Hence, $X_h^*(t) \in \mathbb{D}^{1,2}(\mathbb{L}^2)$. In a similar vein, by [17, Theorem 2.2.2], it follows that

$$\begin{cases} \mathrm{d}D_\mu D_\theta X_h^*(t) = \big[\Delta_h - \mathcal{P}_h(t)\big]D_\mu D_\theta X_h^*(t)\,\mathrm{d}t \\ \qquad\qquad + \beta D_\mu D_\theta X_h^*(t)\,\mathrm{d}W(t) \qquad t \in [\mu\vee\theta, T]\,, \\ D_\mu D_\theta X_h^*(\mu\vee\theta) = \beta D_\mu X_h^*(\theta)\chi_{\{\mu<\theta\}} + \beta D_\theta X_h^*(\mu)\chi_{\{\theta\le\mu\}}\,. \end{cases} \tag{3.60}$$

We may then deduce that $X_h^*(t) \in \mathbb{D}^{2,2}(\mathbb{L}^2)$, and the remaining part of (3.57a).

(4) Verification of (3.57b). Based on the fact that $\mathcal{P}_h(\cdot) \in C\big([0,T]; \mathbb{S}_+(\mathbb{V}_h)\big)$ and Lemma 3.1, it follows that

$$\begin{aligned}&\mathbb{E}\Big[\sup_{t\in[\theta,T]} \|D_\theta X_h^*(t)\|^2_{\dot{\mathbb{H}}_h^1}\Big] + \mathbb{E}\Big[\int_\theta^T \|D_\theta X_h^*(t)\|^2_{\dot{\mathbb{H}}_h^2}\,\mathrm{d}t\Big] \\ &\le C\,\mathbb{E}\big[\|D_\theta X_h^*(\theta)\|^2_{\dot{\mathbb{H}}_h^1}\big] \\ &\le C\big[\|x\|^2_{\mathbb{H}_0^1} + \|\sigma(\cdot)\|^2_{C([0,T];\mathbb{H}_0^1)}\big]\,.\end{aligned}$$

Note that the constant C is independent of θ, which settles the assertion (3.57b).

(5) Verification of (3.58b). Using the linearity of SDE (3.59), the fact that $\mathcal{P}_h(\cdot) \in C\big([0,T];\mathbb{S}_+(\mathbb{V}_h)\big)$, Lemma 3.1, and the triangle inequality, we find that

$$\begin{aligned}
&\mathbb{E}\Big[\sup_{t\in[\theta_1,T]} \|(D_{\theta_1}-D_{\theta_2})X_h^*(t)\|^2\Big]\\
&\le C\,\mathbb{E}\big[\|(D_{\theta_1}-D_{\theta_2})X_h^*(\theta_1)\|^2\big]\\
&\le C\Big[\mathbb{E}\big[\|D_{\theta_1}X_h^*(\theta_1)-D_{\theta_2}X_h^*(\theta_2)\|^2\big]+\mathbb{E}\big[\|D_{\theta_2}X_h^*(\theta_1)-D_{\theta_2}X_h^*(\theta_2)\|^2\big]\Big]\\
&=: C(I_1+I_2)\,.
\end{aligned}$$

For I_1, the result (3.58a) implies that

$$\begin{aligned}
I_2 &\le (\theta_1-\theta_2)\mathbb{E}\Big[\int_0^T \|\Delta_h D_{\theta_2}X_h^*(t)\|^2\,\mathrm{d}t\Big]+\mathbb{E}\Big[\int_{\theta_2}^{\theta_1}\|D_{\theta_2}X_h^*(t)\|^2\,\mathrm{d}t\Big]\\
&\le (\theta_1-\theta_2)\mathbb{E}\Big[\int_{\theta_2}^T \|D_{\theta_2}X_h^*(t)\|^2_{\dot{\mathbb{H}}_h^2}\,\mathrm{d}t+\sup_{t\in[\theta_2,T]}\|D_{\theta_2}X_h^*(t)\|^2\Big]\\
&\le C(\theta_1-\theta_2)\big[\|x\|^2_{\mathbb{H}_0^1}+\|\sigma(\cdot)\|^2_{C([0,T];\mathbb{H}_0^1)}\big]\,.
\end{aligned}$$

For I_2, (3.57a) and (3.57b) lead to

$$\begin{aligned}
I_2 &\le (\theta_1-\theta_2)\mathbb{E}\Big[\int_0^T \|\Delta_h D_{\theta_2}X_h^*(t)\|^2\,\mathrm{d}t\Big]+\mathbb{E}\Big[\int_{\theta_2}^{\theta_1}\|D_{\theta_2}X_h^*(t)\|^2\,\mathrm{d}t\Big]\\
&\le C(\theta_1-\theta_2)\Big[\|x\|^2_{\mathbb{H}_0^1}+\|\sigma(\cdot)\|^2_{C([0,T];\mathbb{H}_0^1)}\Big]\,.
\end{aligned}$$

A combination of the above three estimates now shows the assertion (3.58b). □

The following result is on the regularity of $\big(Y_h(\cdot),Z_h(\cdot)\big)$ of BSDE (3.37).

Lemma 3.8 *Suppose that* $\big(Y_h(\cdot),Z_h(\cdot)\big)$ *solves BSDE* (3.37)*, and* $I_\tau=\{t_n\}_{n=0}^N$ *is a uniform time mesh of* $[0,T]$*. Then there exists a constant* C *independent of* h *such that*

$$\begin{cases}
\mathbb{E}\Big[\sup_{t\in[0,T]}\|Y_h(t)\|^2_{\dot{\mathbb{H}}_h^\gamma}\Big]+\mathbb{E}\Big[\int_0^T \|Y_h(t)\|^2_{\dot{\mathbb{H}}_h^{\gamma+1}}+\|Z_h(t)\|^2_{\dot{\mathbb{H}}_h^\gamma}\,\mathrm{d}t\Big]\\
\qquad\le C\big[\|x\|^2_{\dot{\mathbb{H}}^\gamma}+\|\sigma(\cdot)\|^2_{L^2(0,T;\dot{\mathbb{H}}^\gamma)}\big]\quad \gamma=0,1\,, & (3.61a)\\
\big\|Y_h(\cdot)-Y_h\big(\nu(\cdot)\big)\big\|^2_{L^2_{\mathbb{F}}(0,T;\mathbb{L}^2)}\le C\tau\big[\|x\|^2_{\mathbb{H}_0^1}+\|\sigma\|^2_{L^2(0,T;\mathbb{H}_0^1)}\big], & (3.61b)
\end{cases}$$

and

$$\begin{cases} \sup_{\theta\in[0,T]} \Big\{ \mathbb{E}\Big[\sup_{t\in[\theta,T]} \|D_\theta Y_h(t)\|^2 \Big] + \mathbb{E}\Big[\int_\theta^T \|D_\theta Z_h(t)\|^2 \,\mathrm{d}t \Big] \Big\} \\ \quad + \sup_{\theta,\mu\in[0,T]} \mathbb{E}\Big[\sup_{t\in[\mu\vee\theta,T]} \|D_\mu D_\theta Y_h(t)\|^2 \Big] \\ \quad \le C\big[\|x\|^2 + \|\sigma(\cdot)\|^2_{C([0,T];\mathbb{L}^2)} \big], & (3.62a) \\ \sup_{\theta\in[0,T]} \Big\{ \mathbb{E}\Big[\sup_{t\in[\theta,T]} \|D_\theta Y_h(t)\|^2_{\mathbb{H}^1_h} + \int_\theta^T \|D_\theta Y_h(t)\|^2_{\mathbb{H}^2_h} + \|D_\theta Z_h(t)\|^2_{\mathbb{H}^1_h} \,\mathrm{d}t \Big] \Big\} \\ \quad \le C\big[\|x\|^2_{\mathbb{H}^1_0} + \|\sigma(\cdot)\|^2_{C([0,T];\mathbb{H}^1_0)} \big], & (3.62b) \\ \mathbb{E}\Big[\sup_{t\in[\theta_1,T]} \|(D_{\theta_1} - D_{\theta_2}) Y_h(t)\|^2 \Big] \\ \quad \le C(\theta_1 - \theta_2)\big[\|x\|^2_{\mathbb{H}^1_0} + \|\sigma(\cdot)\|^2_{C([0,T];\mathbb{H}^1_0)} + L^2_{\sigma,1/2} \big] \quad \theta_2 \le \theta_1, & (3.62c) \\ \mathbb{E}\big[\|Z_h(t) - Z_h(s)\|^2 \big] \\ \quad \le C|t-s|\big[\|x\|^2_{\mathbb{H}^1_0} + \|\sigma(\cdot)\|^2_{C([0,T];\mathbb{H}^1_0)} + L^2_{\sigma,1/2} \big] \quad t,s\in[0,T], & (3.62d) \end{cases}$$

where $\nu(\cdot)$ is defined in (2.8).

Proof (1) Verification of (3.61a). This assertion follows from (3.14) in Lemma 3.3 and (3.56a) in Lemma 3.7.

(2) Verification of (3.61b). This assertion follows from (3.15) in Lemma 3.3 and (3.56a) in Lemma 3.7.

(3) Verification of (3.62a), (3.62b). On noting that $X_h^*(t) \in \mathbb{D}^{2,2}(\mathbb{L}^2)$ for any $t \in [0, T]$, by [4, Proposition 5.3] we can get

$$\begin{cases} \mathrm{d}D_\theta Y_h(t) = \big[-\Delta_h D_\theta Y_h(t) - D_\theta Z_h(t) + D_\theta X_h^*(t) \big]\,\mathrm{d}t \\ \qquad\qquad\qquad + D_\theta Z_h(t)\,\mathrm{d}W(t) \quad t\in[\theta,T), \\ D_\theta Y_h(T) = -\alpha D_\theta X_h^*(T), \\ D_\theta Y_h(t) = 0, \;\; D_\theta Z_h(t) = 0 \qquad t\in[0,\theta), \end{cases} \tag{3.63}$$

$$Z_h(t) = D_t Y_h(t) \qquad \text{a.e. } t\in[0,T], \tag{3.64}$$

and

$$\begin{cases} \mathrm{d}D_\mu D_\theta Y_h(t) = \big[-\Delta_h D_\mu D_\theta Y_h(t) - D_\mu D_\theta Z_h(t) + D_\mu D_\theta X_h^*(t) \big]\,\mathrm{d}t \\ \qquad\qquad\qquad + D_\mu D_\theta Z_h(t)\,\mathrm{d}W(t) \qquad t\in[\mu\vee\theta,T), \\ D_\mu D_\theta Y_h(T) = -\alpha D_\mu D_\theta X_h^*(T). \end{cases} \tag{3.65}$$

Then, (3.14) in Lemma 3.3 and ((3.57a)), (3.57b) in Lemma 3.7 lead to (3.62a) and (3.62b).

(4) Verification of (3.62c). Lemma 3.3, the linearity of BSDE (3.37), and (3.58b) in Lemma 3.7 imply

$$\begin{aligned}
&\mathbb{E}\Big[\sup_{t\in[\theta_1,T]}\|(D_{\theta_1}-D_{\theta_2})Y_h(t)\|^2\Big]\\
&\le C\Big\{\mathbb{E}\big[\|(D_{\theta_1}-D_{\theta_2})X_h^*(T)\|^2\big]+\mathbb{E}\Big[\int_{\theta_1}^T\|(D_{\theta_1}-D_{\theta_2})X_h^*(\theta)\|^2\,\mathrm{d}\theta\Big]\Big\}\\
&\le C(\theta_1-\theta_2)\Big[\|x\|^2_{\mathbb{H}_0^1}+\|\sigma(\cdot)\|^2_{C([0,T];\mathbb{H}_0^1)}+L^2_{\sigma,1/2}\Big],
\end{aligned}$$

which settles the assertion (3.62c).

(5) Verification of (3.62d). Suppose that $s\le t$. By applying the fact that $Z_h(\cdot)=D.Y_h(\cdot)$, a.e. and the assertion (3.62c), we arrive at

$$\begin{aligned}
&\mathbb{E}\big[\|Z_h(t)-Z_h(s)\|^2\big]\\
&\le 2\mathbb{E}\big[\|(D_t-D_s)Y_h(t)\|^2\big]+2\mathbb{E}\big[\|D_s\big(Y_h(t)-Y_h(s)\big)\|^2\big]\\
&\le C|t-s|\Big[\|x\|^2_{\mathbb{H}_0^1}+\|\sigma(\cdot)\|^2_{C([0,T];\mathbb{H}_0^1)}+L^2_{\sigma,1/2}\Big]\\
&\quad+2\mathbb{E}\big[\|D_s\big(Y_h(t)-Y_h(s)\big)\|^2\big].
\end{aligned}\tag{3.66}$$

By (3.63) and (3.64), we have

$$\begin{aligned}
&\mathbb{E}\big[\|D_s\big(Y_h(t)-Y_h(s)\big)\|^2\big]\\
&\le C\Big\{\mathbb{E}\Big[\int_s^t\|D_sD_\theta Y_h(\theta)\|^2\,\mathrm{d}\theta\Big]\\
&\quad+(t-s)\mathbb{E}\Big[\int_s^T\|\Delta_hD_sY_h(\theta)\|^2+\|D_sZ_h(\theta)\|^2+\|D_sX_h^*(\theta)\|^2\,\mathrm{d}\theta\Big]\Big\},
\end{aligned}$$

which, together with (3.62a), (3.62b) and (3.57a), leads to

$$\mathbb{E}\big[\|D_s\big(Y_h(t)-Y_h(s)\big)\|^2\big]\le C|t-s|\big[\|x\|^2_{\mathbb{H}_0^1}+\|\sigma(\cdot)\|^2_{C([0,T];\mathbb{H}_0^1)}\big].\tag{3.67}$$

Now (3.62d) can be derived by (3.66)–(3.67). □

Relying on Lemmata 3.1 and 3.8, we can derive the following strong bound for the optimal control $U_h^*(\cdot)$ of Problem **(SLQ)**$_h$.

Lemma 3.9 *Let $U_h^*(\cdot)$ be the optimal control of Problem* **(SLQ)**$_h$. *There exists a constant C such that*

$$\mathbb{E}\Big[\sup_{t\in[0,T]}\|U_h^*(t)\|^2_{\dot{\mathbb{H}}_h^2}\Big]\le C\big[\|x\|^2_{\mathbb{H}_0^1\cap\mathbb{H}^2}+\|\sigma(\cdot)\|^2_{L^2(0,T;\mathbb{H}_0^1\cap\mathbb{H}^2)}\big].$$

Proof By Lemma 3.1 with $\gamma=2$, the optimality condition (3.36) and (3.61a) with $\gamma=0$ in Lemma 3.8, we arrive at

$$
\begin{aligned}
\mathbb{E}\Big[\sup_{t\in[0,T]}\|X_h^*(t)\|_{\dot{\mathbb{H}}_h^2}^2\Big] &\le C\,\mathbb{E}\Big[\,\|X_h^*(0)\|_{\dot{\mathbb{H}}_h^2}^2+\int_0^T \|U_h^*(t)\|_{\dot{\mathbb{H}}_h^1}^2+\|\Pi_h\sigma(t)\|_{\dot{\mathbb{H}}_h^2}^2\,\mathrm{d}t\Big]\\
&\le C\big[\|x\|_{\mathbb{H}_0^1\cap\mathbb{H}^2}^2+\|\sigma(\cdot)\|_{L^2(0,T;\mathbb{H}_0^1\cap\mathbb{H}^2)}^2\big].
\end{aligned}
$$

Subsequently, by using the optimality condition (3.36) and Lemma 3.3 with $\gamma=2$, we derive

$$
\begin{aligned}
\mathbb{E}\Big[\sup_{t\in[0,T]}\|U_h^*(t)\|_{\dot{\mathbb{H}}_h^2}^2\Big] &= \mathbb{E}\Big[\sup_{t\in[0,T]}\|Y_h(t)\|_{\dot{\mathbb{H}}_h^2}^2\Big]\\
&\le C\,\mathbb{E}\Big[\|\alpha X_h^*(T)\|_{\dot{\mathbb{H}}_h^2}^2+\int_0^T \|X_h^*(t)\|_{\dot{\mathbb{H}}_h^1}^2\,\mathrm{d}t\Big]\\
&\le C\big[\|x\|_{\mathbb{H}_0^1\cap\mathbb{H}^2}^2+\|\sigma(\cdot)\|_{L^2(0,T;\mathbb{H}_0^1\cap\mathbb{H}^2)}^2\big].
\end{aligned}
$$

That completes the proof. □

3.3.3 *Temporal Discretization of Problem* (SLQ)$_h$

In this part, we propose a temporal discretization scheme of Problem **(SLQ)**$_h$, consider its well-posedness and derive error estimates. For this purpose, we use a mesh I_τ which covers $[0,T]$, and consider step processes $\big(X_{h\tau}(\cdot),U_{h\tau}(\cdot)\big)\in\mathbb{X}_\mathbb{F}\times\mathbb{U}_\mathbb{F}\subset L^2_\mathbb{F}\big(0,T;\mathbb{V}_h\big)\times L^2_\mathbb{F}\big(0,T;\mathbb{V}_h\big)$, where

$$
\begin{aligned}
\mathbb{X}_\mathbb{F} &\triangleq \big\{X(\cdot)\in L^2_\mathbb{F}(0,T;\mathbb{V}_h)\,\big|\,X(t)=X(t_n),\ \forall t\in[t_n,t_{n+1}),\ n=0,1,\cdots,N-1\big\},\\
\mathbb{U}_\mathbb{F} &\triangleq \big\{U(\cdot)\in L^2_\mathbb{F}(0,T;\mathbb{V}_h)\,\big|\,U(t)=U(t_n),\ \forall t\in[t_n,t_{n+1}),\ n=0,1,\cdots,N-1\big\};
\end{aligned}
$$

for any $X(\cdot)\in\mathbb{X}_\mathbb{F}$ and $U(\cdot)\in\mathbb{U}_\mathbb{F}$, we define[1]

$$
\|X(\cdot)\|_{\mathbb{X}_\mathbb{F}}\triangleq\Big(\tau\sum_{n=0}^{N-1}\mathbb{E}\big[\|X(t_n)\|^2\big]\Big)^{1/2},\quad \|U(\cdot)\|_{\mathbb{U}_\mathbb{F}}\triangleq\Big(\tau\sum_{n=0}^{N-1}\mathbb{E}\big[\|U(t_n)\|^2\big]\Big)^{1/2}.
$$

Now a temporal discretization scheme of Problem **(SLQ)**$_h$ reads as follows.

Problem (SLQ)$_{h\tau}$. Find an optimal control $U_{h\tau}^*(\cdot)\in\mathbb{U}_\mathbb{F}$ which minimizes the quadratic cost functional

$$
\mathcal{J}_{h,\tau}\big(U_{h\tau}(\cdot)\big)=\frac{1}{2}\big[\|X_{h\tau}(\cdot)\|_{\mathbb{X}_\mathbb{F}}^2+\|U_{h\tau}(\cdot)\|_{\mathbb{U}_\mathbb{F}}^2\big]+\frac{\alpha}{2}\mathbb{E}\big[\|X_{h\tau}(T)\|^2\big], \tag{3.68}
$$

subject to the forward semi-implicit difference equation

[1] Another suitable norm for $\mathbb{X}_\mathbb{F}$ could be defined by $\big(\tau\sum_{n=1}^{N}\mathbb{E}\big[\|X(t_n)\|^2\big]\big)^{1/2}$; see [11, 13, 14]. Under this setting, Theorem 3.8 in this section remains valid.

$$\begin{cases} X_{h\tau}(t_{n+1}) - X_{h\tau}(t_n) = \tau\big[\Delta_h X_{h\tau}(t_{n+1}) + U_{h\tau}(t_n)\big] \\ \qquad +\big[\beta X_{h\tau}(t_n) + \Pi_h\sigma(t_n)\big]\Delta_{n+1}W \quad n = 0, 1, \cdots, N-1, \\ X_{h\tau}(0) = \Pi_h x\,. \end{cases} \tag{3.69}$$

The following result states a (discrete) Pontryagin's maximum principle for the uniquely solvable Problem **(SLQ)**$_{h\tau}$.

Theorem 3.7 *Let $A_0 = (\mathbb{1}_h - \tau\Delta_h)^{-1}$. The unique optimal pair $\big(X^*_{h\tau}(\cdot), U^*_{h\tau}(\cdot)\big) \in \mathbb{X}_{\mathbb{F}} \times \mathbb{U}_{\mathbb{F}}$ of Problem* **(SLQ)**$_{h\tau}$ *solves the following coupled equalities for $n = 0, 1, \cdots, N-1$:*

$$\begin{cases} X^*_{h\tau}(t_{n+1}) = A_0^{n+1}\prod_{j=1}^{n+1}(1+\beta\Delta_j W)X_{h\tau}(0) \\ \quad +\sum_{j=0}^{n} A_0^{n+1-j}\prod_{k=j+2}^{n+1}(1+\beta\Delta_k W)\big[\tau U^*_{h\tau}(t_j)+\Pi_h\sigma(t_j)\Delta_{j+1}W\big], & (3.70a) \\ Y_{h\tau}(t_n) = -\tau\mathbb{E}^{t_n}\Big[\sum_{j=n+1}^{N-1} A_0^{j-n}\prod_{k=n+2}^{j}(1+\beta\Delta_k W)X^*_{h\tau}(t_j)\Big] \\ \quad -\alpha\mathbb{E}^{t_n}\Big[A_0^{N-n}\prod_{k=n+2}^{N}(1+\beta\Delta_k W)X^*_{h\tau}(T)\Big], & (3.70b) \\ X^*_{h\tau}(0) = \Pi_h x\,, & (3.70c) \end{cases}$$

together with the discrete optimality condition

$$U^*_{h\tau}(t_n) - Y_{h\tau}(t_n) = 0\,. \tag{3.71}$$

Below, we write $Y_{h\tau}(\cdot; X^*_{h\tau})$ for $Y_{h\tau}(\cdot)$ to indicate the dependence on $X^*_{h\tau}(\cdot)$.

We mention the appearing mapping $Y_{h\tau}(\cdot)$ in (3.70b), which is used to give the discrete optimality condition (3.71); see also Remark 3.5. We will use this characterization in Theorem 3.7 of $\big(X^*_{h\tau}(\cdot), U^*_{h\tau}(\cdot)\big) \in \mathbb{X}_{\mathbb{F}} \times \mathbb{U}_{\mathbb{F}}$ to deduce rates of convergence for the optimal pair of Problem **(SLQ)**$_{h\tau}$ in the next Theorem 3.8; but to actually implement this numerical method is still too ambitious: this is the reason why another method that is suitable for implementation will be proposed in the subsequent Sects. 3.3.4 and 3.4.

Theorem 3.8 *Suppose that* **(A)** *holds. Let $\big(X^*_h(\cdot), U^*_h(\cdot)\big)$ be the optimal pair to Problem* **(SLQ)**$_h$, *and $\big(X^*_{h\tau}(\cdot), U^*_{h\tau}(\cdot)\big)$ solves Problem* **(SLQ)**$_{h\tau}$. *Then, there exists a constant C independent of h, τ such that*

$$\begin{aligned} &\sum_{n=0}^{N-1}\mathbb{E}\Big[\int_{t_n}^{t_{n+1}} \|U^*_h(t) - U^*_{h\tau}(t_n)\|^2 + \|X^*_h(t) - X^*_{h\tau}(t_n)\|^2\,\mathrm{d}t\Big] \\ &\quad \le C\tau\big[\|x\|^2_{\mathbb{H}^1_0\cap\mathbb{H}^2} + \|\sigma(\cdot)\|^2_{C([0,T];\mathbb{H}^1_0)\cap L^2(0,T;\mathbb{H}^1_0\cap\mathbb{H}^2)} + L^2_{\sigma,1/2}\big], \end{aligned} \tag{3.72}$$

and

$$\max_{0\le n\le N}\mathbb{E}\big[\|X_h^*(t_n)-X_{h\tau}^*(t_n)\|^2\big]+\tau\sum_{n=0}^{N-1}\mathbb{E}\big[\|X_h^*(t_n)-X_{h\tau}^*(t_n)\|_{\mathbb{H}_0^1}^2\big]$$
$$\le C\tau\big[\|x\|_{\mathbb{H}_0^1\cap\mathbb{H}^2}^2+\|\sigma(\cdot)\|_{C([0,T];\mathbb{H}_0^1)\cap L^2(0,T;\mathbb{H}_0^1\cap\mathbb{H}^2)}^2+L_{\sigma,1/2}^2\big]. \tag{3.73}$$

Remark 3.5 **(i)** Equation (3.70a) is the time-implicit approximation of SDE (3.35) with $U_{h\tau}(\cdot)=U_{h\tau}^*(\cdot)$, while $Y_{h\tau}(\cdot)$ is an approximation of $Y_h(\cdot)$ to BSDE (3.37); in fact, (3.70b) is different from the temporal discretization of (3.37) via the implicit Euler method—which is why our approach differs from the one known as 'first optimize, then discretize'. See also Lemma 3.10 for an (auxiliary, non-implementable) approximation of $Y_h(\cdot)$ that is based on the implicit Euler method. In Lemma 3.11, we estimate the difference of these two approximations.

(ii) If we endow the discrete state space with the following norm $\|X(\cdot)\|_{\mathbb{X}_\mathbb{F}}\triangleq\big(\tau\sum_{n=1}^N\mathbb{E}\big[\|X(t_n)\|^2\big]\big)^{1/2}$, then (3.70b) turns to

$$Y_{h\tau}(t_n)=-\tau\mathbb{E}^{t_n}\Big[\sum_{j=n+1}^{N}A_0^{j-n}\prod_{k=n+2}^{j}(1+\beta\Delta_kW)X_{h\tau}^*(t_j)\Big]$$
$$-\alpha\mathbb{E}^{t_n}\Big[A_0^{N-n}\prod_{k=n+2}^{N}(1+\beta\Delta_kW)X_{h\tau}^*(T)\Big]; \tag{3.74}$$

see [14, Theorem 3.2]. Hence, when SPDE (1.2) is driven by *additive* noise only (i.e., $\beta=0$), $Y_{h\tau}(\cdot)$ in (3.74) satisfies

$$\begin{cases}[\mathbb{1}_h-\tau\Delta_h]Y_{h\tau}(t_n)=\mathbb{E}^{t_n}\big[Y_{h\tau}(t_{n+1})-\tau X_{h\tau}^*(t_{n+1})\big] & n=0,1,\cdots,N-1,\\ Y_{h\tau}(T)=-\alpha X_{h\tau}^*(T),\end{cases}$$

which is the same as the semi-implicit Euler method; see Lemma 3.10 and [11, Theorem 4.2]. In this case, to 'first discretize, then optimize', and to 'first optimize, then discretize' lead to the same discrete optimality system.

Proof of Theorem 3.7. We divide the proof into three steps.

(1) Recall the notation $A_0=(\mathbb{1}_h-\tau\Delta_h)^{-1}$. We define the bounded operators $\Gamma:\mathbb{V}_h\to\mathbb{X}_\mathbb{F}$ and $L:\mathbb{U}_\mathbb{F}\to\mathbb{X}_\mathbb{F}$ as follows, for any $n=0,1,\cdots,N$,

$$\big(\Gamma X_{h\tau}(0)\big)(t_n)=A_0^n\prod_{j=1}^{n}\big(1+\beta\Delta_jW\big)X_{h\tau}(0),$$

$$\text{and}\quad \big(LU_{h\tau}(\cdot)\big)(t_n)=\tau\sum_{j=0}^{n-1}A_0^{n-j}\prod_{k=j+2}^{n}(1+\beta\Delta_kW)\,U_{h\tau}(t_j), \tag{3.75}$$

respectively, where $\big(X_{h\tau}(\cdot), U_{h\tau}(\cdot)\big)$ is an admissible pair of Problem **(SLQ)**$_{h\tau}$; see (3.68)–(3.69). We also need $f(\cdot)$, which we define as

$$f(t_n) = \sum_{j=0}^{n-1} A_0^{n-j} \prod_{k=j+2}^{n} (1+\beta\Delta_k W)\, \Pi_h \sigma(t_j)\Delta_{j+1}W \qquad \forall n = 0, 1, \cdots, N\,,$$

and use below the abbreviations

$$\widehat{\Gamma}\Pi_h x \triangleq (\Gamma\Pi_h x)(T)\,, \qquad \widehat{L}U_{h\tau}(\cdot) \triangleq \big(LU_{h\tau}(\cdot)\big)(T)\,, \qquad \widehat{f} \triangleq f(T)\,. \tag{3.76}$$

By (3.69), we find that

$$X_{h\tau}(t_n) = \big(\Gamma X_{h\tau}(0)\big)(t_n) + \big(LU_{h\tau}(\cdot)\big)(t_n) + f(t_n) \qquad n = 0, 1, \cdots, N\,. \tag{3.77}$$

It is easy to check that, for any $\xi(\cdot) \in \mathbb{X}_{\mathbb{F}}$, and any $\eta \in L^2_{\mathcal{F}_T}(\Omega; \mathbb{V}_h)$,

$$\big(L^*\xi(\cdot)\big)(t_n) = \mathbb{E}^{t_n}\Big[\tau \sum_{j=n+1}^{N-1} A_0^{j-n} \prod_{k=n+2}^{j} (1+\beta\Delta_k W)\xi(t_j)\Big]\,,$$

$$\text{and} \quad \big(\widehat{L}^*\eta\big)(t_n) = \mathbb{E}^{t_n}\Big[A_0^{N-n} \prod_{k=n+2}^{N} (1+\beta\Delta_k W)\eta\Big] \quad n = 0, 1, \cdots, N-1\,.$$

(2) By (3.75)–(3.77), we can rewrite $\mathcal{J}_{h,\tau}\big(U_{h\tau}(\cdot)\big)$ in (3.68) as follows:

$$\begin{aligned}
&\mathcal{J}_{h,\tau}\big(U_{h\tau}(\cdot)\big)\\
&= \frac{1}{2}\Big\{\Big(\big([\mathbb{1}_h + L^*L + \alpha\widehat{L}^*\widehat{L}]U_{h\tau}(\cdot)\big)(\cdot), U_{h\tau}(\cdot)\Big)_{L^2_{\mathbb{F}}(0,T;\mathbb{L}^2)}\\
&\quad +2\Big(\big([L^*\Gamma + \alpha\widehat{L}^*\widehat{\Gamma}]\Pi_h x\big)(\cdot) + \big(L^* f(\cdot)\big)(\cdot) + \alpha\big(\widehat{L}^*\widehat{f}\big)(\cdot), U_{h\tau}(\cdot)\Big)_{L^2_{\mathbb{F}}(0,T;\mathbb{L}^2)}\\
&\quad +\Big[\big((\Gamma\Pi_h x)(\cdot) + f(\cdot), (\Gamma\Pi_h x)(\cdot) + f(\cdot)\big)_{L^2_{\mathbb{F}}(0,T;\mathbb{L}^2)}\\
&\qquad +\alpha\big(\widehat{\Gamma}\Pi_h x + \widehat{f}, \widehat{\Gamma}\Pi_h x + \widehat{f}\big)_{L^2_{\mathcal{F}_T}(\Omega;\mathbb{L}^2)}\Big]\Big\}\\
&=: \frac{1}{2}\Big[\Big(\big(\mathfrak{N}U_{h\tau}(\cdot)\big)(\cdot), U_{h\tau}(\cdot)\Big)_{L^2_{\mathbb{F}}(0,T;\mathbb{L}^2)} + 2\big(\mathfrak{H}(\Pi_h x, f)(\cdot), U_{h\tau}(\cdot)\big)_{L^2_{\mathbb{F}}(0,T;\mathbb{L}^2)}\\
&\quad +\mathfrak{M}(\Pi_h x, f)\Big]\,.
\end{aligned}$$

We use this re-writing of $\mathcal{J}_{h,\tau}\big(U_{h\tau}(\cdot)\big)$ which involves mappings $\mathfrak{N}$ and $\mathfrak{H}$, and note that the last term does not depend on $U_{h\tau}(\cdot)$, to now involve the optimality condition for $U^*_{h\tau}(\cdot)$: since $\mathfrak{N} = \mathbb{1}_h + L^*L + \alpha\widehat{L}^*\widehat{L}$ is positive definite, there exists a unique $U^*_{h\tau}(\cdot) \in \mathbb{U}_{\mathbb{F}}$ such that

$$\big(\mathfrak{N}U^*_{h\tau}(\cdot)\big)(\cdot)+\mathfrak{H}(\Pi_h x,\, f)(\cdot)=0\,. \tag{3.78}$$

Therefore, for any $U_{h\tau}(\cdot)\in\mathbb{U}_{\mathbb{F}}$ such that $U_{h\tau}(\cdot)\neq U^*_{h\tau}(\cdot)$,

$$\begin{aligned}&\mathcal{J}_{h,\tau}\big(U_{h\tau}(\cdot)\big)-\mathcal{J}_{h,\tau}\big(U^*_{h\tau}(\cdot)\big)\\&\quad=\frac{1}{2}\Big(\mathfrak{N}\big(U_{h\tau}(\cdot)-U^*_{h\tau}(\cdot)\big)(\cdot),\,U_{h\tau}(\cdot)-U^*_{h\tau}(\cdot)\Big)_{L^2_{\mathbb{F}}(0,T;\mathbb{L}^2)}>0\,,\end{aligned}$$

which means that $U^*_{h\tau}(\cdot)$ is the unique optimal control, and $\big(X^*_{h\tau}(\cdot),\,U^*_{h\tau}(\cdot)\big)$ is the unique optimal pair.

(3) By the definition of $\mathfrak{N}$, $\mathfrak{H}$, and (3.77), we deduce for (3.78) that

$$0=\big(\mathfrak{N}U^*_{h\tau}(\cdot)\big)(\cdot)+\mathfrak{H}(\Pi_h x,\, f)(\cdot)=U^*_{h\tau}(\cdot)+\big(L^*X^*_{h\tau}(\cdot)+\alpha\widehat{L}^*X^*_{h\tau}(T)\big)(\cdot)\,.$$

For $Y_{h\tau}(\cdot)$ in (3.70b), we compute

$$\big(L^*X^*_{h\tau}(\cdot)+\alpha\widehat{L}^*X^*_{h\tau}(T)\big)(\cdot)=-Y_{h\tau}(\cdot)\,. \tag{3.79}$$

Then (3.71) can be deduced by these two equalities. That completes the proof.

Our next goal is to verify Theorem 3.8 which needs preparatory results that are stated in Lemmata 3.10 through 3.13.

In Remark 3.5 we have already mentioned that $Y_{h\tau}(\cdot)$ in (3.70b) neither solves the temporal discretization of BSDE (3.37) by the explicit Euler, nor by the implicit Euler method in general. The following two lemmata study the difference between $Y_{h\tau}(\cdot)$ and the temporal discretization of BSDE (3.37) by the implicit Euler method.

Lemma 3.10 *Suppose that $\big(Y_\cdot,\ Z_0(\cdot)\big)$ solves the following backward equation:*

$$\begin{cases}Y_n-Y_{n+1}=\tau\Delta_h Y_n+\displaystyle\int_{t_n}^{t_{n+1}}\beta\bar{Z}_0(t)-X^*_{h\tau}\big(\mu(t)\big)\,\mathrm{d}t-\int_{t_n}^{t_{n+1}}Z_0(t)\,\mathrm{d}W(t)\,,\\ Y_N=-\alpha X^*_{h\tau}(T)\,,\end{cases} \tag{3.80}$$

where $X^*_{h\tau}$ is the optimal state of Problem **(SLQ)**$_{h\tau}$*, $\mu(\cdot)$ is defined in (2.8) and $\bar{Z}_0(\cdot)$ is a piecewise constant process which is defined by*

$$\bar{Z}_0(t)=\frac{1}{\tau}\mathbb{E}^{t_n}\Big[\int_{t_n}^{t_{n+1}}Z_0(s)\,\mathrm{d}s\Big]\quad\forall\, t\in[t_n,t_{n+1})\,,\qquad n=0,1,\cdots,N-1\,.$$

Then,

$$\begin{aligned}Y_n=&-\tau\mathbb{E}^{t_n}\Big[\sum_{j=n+1}^{N}A_0^{j-n}\prod_{k=n+1}^{j}(1+\beta\Delta_k W)\,X^*_{h\tau}(t_j)\Big]\\&-\alpha\mathbb{E}^{t_n}\Big[A_0^{N-n}\prod_{k=n+1}^{N}(1+\beta\Delta_k W)\,X^*_{h\tau}(T)\Big]\qquad n=0,1,\cdots,N-1\,.\end{aligned}$$

Proof The well-posedness of (3.80) follows from Lemma 3.4. For any $n = 0, 1, \cdots, N-1$, it is easy to see that

$$\int_{t_n}^{t_{n+1}} Z_0(t)\, \mathrm{d}t = \mathbb{E}^{t_n}\Big[\big(Y_{n+1} - \tau X_{h\tau}^*(t_{n+1})\big)\Delta_{n+1} W\Big],$$

which yields the desired result. □

The solution $\big(Y_\cdot,\ Z_0(\cdot)\big)$ in (3.80) depends on $X_{h\tau}^*$. Hence, in what follows, we may write it in the form $\big(Y_\cdot(X_{h\tau}^*),\ Z_0(\cdot;\ X_{h\tau}^*)\big)$.

Similar to $\mathcal{S}$ and $\mathcal{S}_h$, we can define the solution operator for the difference equation (3.69), $\mathcal{S}_{h\tau} : \mathbb{U}_{\mathbb{F}} \to \mathbb{X}_{\mathbb{F}}$. In the next lemma, we estimate the difference between $Y_{h\tau}\big(\cdot;\ \mathcal{S}_{h\tau}(\Pi_\tau U_h^*)\big)$, which was introduced in (3.70b), and $Y_\cdot\big(\mathcal{S}_{h\tau}(\Pi_\tau U_h^*)\big)$ from Lemma 3.10, which is crucial in proving rates of convergence for the temporal discretization of Problem $\textbf{(SLQ)}_h$.

Lemma 3.11 *Suppose that* $\big(Y_\cdot,\ Z_0(\cdot)\big)$ *solves* (3.80). *Then it holds that*

$$\begin{aligned}\max_{0 \le n \le N-1} \mathbb{E}\Big[\big\| Y_n\big(\mathcal{S}_{h\tau}(\Pi_\tau U_h^*)\big) - Y_{h\tau}\big(t_n;\ \mathcal{S}_{h\tau}(\Pi_\tau U_h^*)\big)\big\|^2\Big] \\ \le C\tau\Big[\|x\|^2 + \|\sigma(\cdot)\|^2_{C([0,T];\mathbb{L}^2)}\Big],\end{aligned}$$

where Π_τ *is defined in* P(2.9), $Y_{h\tau}(\cdot)$ *is given by* (3.70b), *and the constant* C *is independent of* h *and* τ.

Proof By Lemma 3.10 and Theorem 3.7, we find that for any $n = 0, 1, \cdots, N-1$

$$\begin{aligned}&Y_n\big(\mathcal{S}_{h\tau}(\Pi_\tau U_h^*)\big) - Y_{h\tau}\big(t_n;\ \mathcal{S}_{h\tau}(\Pi_\tau U_h^*)\big) \\ &= \Big\{ -\alpha\beta\mathbb{E}^{t_n}\Big[A_0^{N-n}\Delta_{n+1}W \prod_{k=n+2}^{N}(1+\beta\Delta_k W)\mathcal{S}_{h\tau}(T;\ \Pi_\tau U_h^*)\Big] \\ &\quad -\tau\mathbb{E}^{t_n}\Big[A_0^{N-n}\prod_{i=n+2}^{N}(1+\beta\Delta_i W)\,\mathcal{S}_{h\tau}(T;\ \Pi_\tau U_h^*)\Big]\Big\} \\ &\quad -\tau\beta\sum_{k=n+1}^{N}\mathbb{E}^{t_n}\Big[A_0^{k-n}\Delta_{n+1}W\prod_{i=n+2}^{k}(1+\beta\Delta_i W)\,\mathcal{S}_{h\tau}(t_k;\ \Pi_\tau U_h^*)\Big] \\ &=: I_n^0 - \tau\beta\sum_{k=n+1}^{N} I_n^k.\end{aligned}$$

For I_n^k, by (3.70a), we have

$$
\begin{aligned}
I_n^k &= \mathbb{E}^{t_n}\Big[A_0^{k-n}\Delta_{n+1}W\prod_{i=n+2}^{k}(1+\beta\Delta_i W)\,A_0^k\prod_{l=1}^{k}(1+\beta\Delta_l W)\,X_{h\tau}(0)\Big]\\
&\quad+\mathbb{E}^{t_n}\Big[\tau A_0^{k-n}\Delta_{n+1}W\prod_{i=n+2}^{k}(1+\beta\Delta_i W)\sum_{l=0}^{k-1}A_0^{k-l}\prod_{m=l+2}^{k}(1+\beta\Delta_m W)\,U_h^*(t_l)\Big]\\
&\quad+\mathbb{E}^{t_n}\Big[A_0^{k-n}\Delta_{n+1}W\prod_{i=n+2}^{k}(1+\beta\Delta_i W)\sum_{l=0}^{k-1}A_0^{k-l}\\
&\qquad\times\prod_{m=l+2}^{k}(1+\beta\Delta_m W)\,\Delta_{l+1}W\,\Pi_h\sigma(t_l)\Big]\\
&=:\sum_{i=1}^{3}I_{ni}^k\,.
\end{aligned}
\tag{3.81}
$$

By the mutual independence of $\{\Delta_n W\}_{n=1}^N$, and the fact that $X_{h\tau}(0)$ is deterministic, we obtain

$$
\begin{aligned}
\mathbb{E}\big[\|I_{n1}^k\|^2\big] &\le \mathbb{E}\Big[\Big\|\prod_{i=n+2}^{k}(1+\Delta_i W)\,\Delta_{n+1}W\prod_{\ell=1}^{k}(1+\Delta_\ell W)\,X_{h\tau}(0)\Big\|^2\Big]\\
&=\prod_{i=n+2}^{k}\mathbb{E}\big[(1+\Delta_i W)^4\big]\mathbb{E}\big\{\big[\Delta_{n+1}W+\big(\Delta_{n+1}W\big)^2\big]^2\big\}\\
&\qquad\times\prod_{\ell=1}^{n}\mathbb{E}\big[(1+\Delta_\ell W)^2\big]\,\|X_{h\tau}(0)\|^2\\
&\le C\tau\|X_{h\tau}(0)\|\,.
\end{aligned}
\tag{3.82}
$$

For I_{n2}^k, firstly we have

$$
\mathbb{E}\big[\|I_{n2}^k\|^2\big]\le C\tau\sum_{l=0}^{k-1}\mathbb{E}\Big[\Big\|\prod_{i=n+2}^{k}(1+\beta\Delta_i W)\,\Delta_{n+1}W\prod_{m=l+2}^{k}(1+\beta\Delta_m W)\,U_h^*(t_l)\Big\|^2\Big].
$$

Now we divide this estimate into two cases.

Case (i) $l\le n$. In this case, by using the same trick as that to derive (3.82), we can see that

$$
\begin{aligned}
&\mathbb{E}\Big[\Big\|\prod_{i=n+2}^{k}(1+\beta\Delta_i W)\,\Delta_{n+1}W\prod_{m=l+2}^{k}(1+\beta\Delta_m W)\,U_h^*(t_l)\Big\|^2\Big]\\
&\le\prod_{i=n+2}^{k}\mathbb{E}\big[(1+\beta\Delta_i W)^4\big]\mathbb{E}\big[\big(\Delta_{n+1}W+\beta(\Delta_{n+1}W)^2\big)^2\big]
\end{aligned}
$$

$$\times \prod_{m=l+2}^{n} \mathbb{E}\big[(1+\beta\Delta_m W)^2\big]\mathbb{E}\big[\|U_h^*(t_l)\|^2\big]$$
$$\le C\tau \sup_{t\in[0,T]} \mathbb{E}\big[\|U_h^*(t)\|^2\big]. \tag{3.83}$$

Case (ii) $l > n$. Still applying the mutual independence of $\{\Delta_n W\}_{n=1}^N$, we have

$$\mathbb{E}\Big[\Big\| \prod_{i=n+2}^{k} (1+\beta\Delta_i W)\,\Delta_{n+1}W \prod_{m=l+2}^{k} (1+\beta\Delta_m W)\,U_h^*(t_l)\Big\|^2\Big]$$
$$\le \prod_{i=l+2}^{k} \mathbb{E}\big[(1+\beta\Delta_i W)^4\big]\mathbb{E}\big[(1+\beta\Delta_{l+1}W)^2\big]$$
$$\times\Big\{ \prod_{m=n+2}^{l} \mathbb{E}\big[(1+\beta\Delta_m W)^4\big]\mathbb{E}\big[(\Delta_{n+1}W)^4\big]\mathbb{E}\big[\|U_h^*(t_l)\|^4\big]\Big\}^{1/2}$$
$$\le C\tau\Big(\sup_{t\in[0,T]} \mathbb{E}\big[\|U_h^*(t)\|^4\big]\Big)^{1/2}. \tag{3.84}$$

We may now combine both estimates and use Lemma 3.7 to conclude that

$$\mathbb{E}\big[\|I_{n2}^k\|^2\big] \le C\tau\Big[\sup_{t\in[0,T]} \mathbb{E}\big[\|U_h^*(t)\|^2\big] + \Big(\sup_{t\in[0,T]} \mathbb{E}\big[\|U_h^*(t)\|^4\big]\Big)^{1/2}\Big]$$
$$\le C\tau\big[\|x\|^2 + \|\sigma(\cdot)\|^2_{C([0,T];\mathbb{L}^2)}\big].$$

To estimate $\mathbb{E}\big[\|I_{n3}^k\|^2\big]$, we need a longer computation. By introducing a notation

$$a_l \triangleq A_0^{2k-n-l}\Delta_{n+1}W \prod_{i=n+2}^{k} (1+\beta\Delta_i W) \prod_{m=l+2}^{k} (1+\beta\Delta_m W)\,\Delta_{l+1}W\,\Pi_h\sigma(t_l),$$

we find that $I_{n3}^k = \sum_{l=0}^{k-1} \mathbb{E}^{t_n}[a_l]$, and divide the summation into three cases.
Case (i) $l \le n-1$. We use the facts that $\{\Delta_m W\}_{m\le n}$ is $\mathcal{F}_{t_n}$-measurable and the mutual independence of $\{\Delta_i W\}_{i=n+1}^k$ to conclude that

$$\mathbb{E}^{t_n}\big[a_l\big] = A_0^{2k-n-l}\Pi_h\sigma(t_l)\Delta_{l+1}W \prod_{m=l+2}^{n} (1+\beta\Delta_m W)$$
$$\times\mathbb{E}^{t_n}\big[(1+\beta\Delta_{n+1}W)\Delta_{n+1}W\big] \prod_{i=n+2}^{k} \mathbb{E}^{t_n}\big[(1+\beta\Delta_i W)^2\big]$$
$$= \beta\tau A_0^{2k-n-l}\Pi_h\sigma(t_l)\Delta_{l+1}W \prod_{m=l+2}^{n} (1+\beta\Delta_m W) \prod_{i=n+2}^{k} (1+\beta^2\tau),$$

and by the fact that $\|A_0\|_{\mathcal{L}(\mathbb{L}^2|_{\mathbb{V}_h})} \leq 1$ we arrive at

$$\begin{aligned}
&\mathbb{E}\big[\big\|\mathbb{E}^{t_n}\big[a_l\big]\big\|^2\big] \\
&\leq C\tau^2\|\sigma(t_l)\|^2\mathbb{E}\big[\Delta_{l+1}^2 W\big] \prod_{m=l+2}^{n} \mathbb{E}\big[(1+\beta\Delta_m W)^2\big] \prod_{i=n+2}^{k} (1+\beta^2\tau)^2 \\
&\leq C\tau^3\|\sigma(t_l)\|^2 .
\end{aligned}$$

Case (ii) $l = n$. Similar to the above case, it follows that

$$\mathbb{E}^{t_n}\big[a_n\big] = A_0^{2k-2n}\Pi_h\sigma(t_n)\tau \prod_{i=n+2}^{k} (1+\beta^2\tau),$$

and

$$\mathbb{E}\big[\big\|\mathbb{E}^{t_n}\big[a_n\big]\big\|^2\big] \leq C\tau^2\|\sigma(t_n)\|^2 \prod_{i=n+2}^{k} (1+\beta^2\tau)^2 \leq C\tau^2\|\sigma(t_l)\|^2 .$$

Case (iii) $l \geq n+1$. Based on the mutual independence of $\{\Delta_i W\}_{i=n+1}^k$ and the fact that $\mathbb{E}\big[\Delta_{n+1}W\big] = 0$, we have

$$\begin{aligned}
\mathbb{E}^{t_n}\big[a_l\big] = A_0^{2k-n-l}&\Pi_h\sigma(t_l)\mathbb{E}\big[\Delta_{n+1}W\big] \prod_{i=n+2}^{l} \mathbb{E}\big[(1+\beta\Delta_i W)\big] \\
&\times\mathbb{E}\big[(1+\beta\Delta_{l+1}W)\Delta_{l+1}W\big] \prod_{m=l+2}^{k} \mathbb{E}\big[(1+\beta\Delta_m W)^2\big] \\
= 0 .&
\end{aligned}$$

Combining with the above estimates, we conclude that

$$\begin{aligned}
\mathbb{E}\big[\|I_{n3}^k\|^2\big] &= \mathbb{E}\Big[\Big\|\sum_{l=0}^{n-1}\mathbb{E}^{t_n}\big[a_l\big] + \mathbb{E}^{t_n}\big[a_n\big] + \sum_{l=n+1}^{k-1}\mathbb{E}^{t_n}\big[a_l\big]\Big\|^2\Big] \\
&\leq C\Big\{n\sum_{l=0}^{n-1}\mathbb{E}\big[\big\|\mathbb{E}^{t_n}\big[a_l\big]\big\|^2\big] + \mathbb{E}\big[\big\|\mathbb{E}^{t_n}\big[a_n\big]\big\|^2\big]\Big\} \\
&\leq C\tau\|\sigma(\cdot)\|_{C([0,T];\mathbb{L}^2)}^2 .
\end{aligned}$$

Note that the constant C in the above estimates is independent of k and n. Finally, by applying these estimates in (3.81), we deduce that

$$
\begin{aligned}
\mathbb{E}\Big[\Big\|\tau\beta\sum_{k=n+1}^{N} I_n^k\Big\|^2\Big] &\leq C\tau^2(N-n)\sum_{k=n+1}^{N}\sum_{i=1}^{3}\mathbb{E}\big[\big\|I_{ni}^k\big\|^2\big] \\
&\leq C\max_{0\leq n\leq N-1}\max_{n+1\leq k\leq N}\sum_{i=1}^{3}\mathbb{E}\big[\big\|I_{ni}^k\big\|^2\big] \\
&\leq C\tau\big[\|x\|^2+\|\sigma(\cdot)\|^2_{C([0,T];\mathbb{L}^2)}\big]. \qquad (3.85)
\end{aligned}
$$

For the term I_n^0, in the same vein, we can also derive

$$
\mathbb{E}\big[\|I_n^0\|^2\big]\leq C\tau\big[\|x\|^2+\|\sigma(\cdot)\|^2_{C([0,T];\mathbb{L}^2)}\big],
$$

which, together with (3.85), yields the assertion. □

In the following, our goal is to estimate the difference between $Y_h\big(\cdot;\mathcal{S}_{h\tau}(\Pi_\tau U_h^*)\big)$ and $Y_\cdot\big(\mathcal{S}_{h\tau}(\Pi_\tau U_h^*)\big)$. To accomplish it, we need to estimate $Z_h\big(\cdot;\mathcal{S}_{h\tau}(\Pi_\tau U_h^*)\big)$.

Lemma 3.12 *Let* $\big(Y_h\big(\cdot;\mathcal{S}_{h\tau}(\Pi_\tau U_h^*)\big), Z_h\big(\cdot;\mathcal{S}_{h\tau}(\Pi_\tau U_h^*)\big)\big)$ *solve* (3.37) *with* $X_h^*(\cdot)=\mathcal{S}_{h\tau}(\cdot;\Pi_\tau U_h^*)$. *There exists a constant* C *independent of* h,τ *such that*

$$
\begin{aligned}
&\mathbb{E}\Big[\big\|Z_h\big(t;\mathcal{S}_{h\tau}(\Pi_\tau U_h^*)\big)-Z_h\big(\nu(t);\mathcal{S}_{h\tau}(\Pi_\tau U_h^*)\big)\big\|^2\Big] \\
&\quad\leq C|t-\nu(t)|\big[\|x\|^2_{\mathbb{H}_0^1}+\|\sigma(\cdot)\|^2_{C([0,T];\mathbb{H}_0^1)}+L^2_{\sigma,1/2}\big]\qquad\forall t\in[0,T],
\end{aligned}
\tag{3.86}
$$

where $\nu(\cdot)$ *is defined in* (2.8).

Proof The proof consists of three steps.

(1) We claim that $\mathcal{S}_{h\tau}(t_n;\Pi_\tau U_h^*)\in\mathbb{D}^{2,2}(\mathbb{L}^2)$, for any $n=0,1,\cdots,N$. Indeed, by (3.77) we know that

$$
\begin{aligned}
&\mathcal{S}_{h\tau}(t_n;\Pi_\tau U_h^*) \\
&=A_0^n\prod_{j=1}^{n}(1+\beta\Delta_jW)X_{h\tau}(0)+\tau\sum_{j=0}^{n-1}A_0^{n-j}\prod_{k=j+2}^{n}(1+\beta\Delta_kW)U_h^*(t_j) \\
&\quad+\sum_{j=0}^{n-1}A_0^{n-j}\prod_{k=j+2}^{n}(1+\beta\Delta_kW)\Pi_h\sigma(t_j)\Delta_{j+1}W.
\end{aligned}
$$

In the following, we only prove that the second term on the right-hand side of the above representation is in $\mathbb{D}^{2,2}(\mathbb{L}^2)$. The other two terms can be proved in a similar vein.

By the optimality condition (3.36) and Lemma 3.8, we know that $U_h^*(t_j)\in\mathbb{D}^{2,2}(\mathbb{L}^2)$, for any $j=0,1,\cdots,N-1$. For any $\theta,\mu\in[0,T]$, without loss of generality, suppose that $\theta\in[t_l,t_{l+1})$, $\mu\in[t_m,t_{m+1})$. Then, by the chain rule and the fact that $D_\theta(1+\beta\Delta_k W) = \beta\delta_{lk}$, $D_\mu(1+\beta\Delta_k W) = \beta\delta_{mk}$, where δ_{lk}, δ_{mk} are Kronecker delta functions, we have

$$\begin{aligned} & D_\theta\Big(\tau\sum_{j=0}^{n-1} A_0^{n-j} \prod_{k=j+2}^{n} (1+\beta\Delta_k W)\, U_h^*(t_j)\Big) \\ &= \tau\sum_{j=0}^{n-1} \beta\chi_{\{j+2\le l\le n\}} A_0^{n-j} \prod_{\substack{k=j+2\\ k\neq l}}^{n} (1+\beta\Delta_k W) U_h^*(t_j) \\ &+\tau\sum_{j=0}^{n-1} A_0^{n-j} \prod_{k=j+2}^{n} (1+\beta\Delta_k W) D_\theta U_h^*(t_j)\,, \end{aligned}$$

and

$$\begin{aligned} & D_\mu D_\theta\Big(\tau\sum_{j=0}^{n-1} A_0^{n-j} \prod_{k=j+2}^{n} (1+\beta\Delta_k W)\, U_h^*(t_j)\Big) \\ &= \tau\sum_{j=0}^{n-1} \beta^2\chi_{\{j+2\le l\le n\}}\chi_{\{j+2\le m\le n\}}(1-\delta_{ml}) A_0^{n-j} \prod_{\substack{k=j+2\\ k\neq l,\, k\neq m}}^{n} (1+\beta\Delta_k W)\, U_h^*(t_j) \\ &+\tau\sum_{j=0}^{n-1} \beta\chi_{\{j+2\le l\le n\}} A_0^{n-j} \prod_{\substack{k=j+2\\ k\neq l}}^{n} (1+\beta\Delta_k W)\, D_\mu U_h^*(t_j) \\ &+\tau\sum_{j=0}^{n-1} \chi_{\{j+2\le m\le n\}} A_0^{n-j} \prod_{\substack{k=j+2\\ k\neq m}}^{n} (1+\beta\Delta_k W)\, D_\theta U_h^*(t_j) \\ &+\tau\sum_{j=0}^{n-1} A_0^{n-j} \prod_{k=j+2}^{n} (1+\beta\Delta_k W)\, D_\mu D_\theta U_h^*(t_j)\,, \end{aligned}$$

and subsequently, by (3.61a) and (3.62a) in Lemma 3.8,

$$\begin{aligned} & \mathbb{E}\Big[\Big\| D_\mu D_\theta\Big(\tau\sum_{j=0}^{n-1} A_0^{n-j} \prod_{k=j+2}^{n} (1+\Delta_k W)\, U_h^*(t_j)\Big)\Big\|^2\Big] \\ &\le C \max_{0\le j\le N-1} \mathbb{E}\Big[\|U_h^*(t_j)\|^2 + \|D_\theta U_h^*(t_j)\|^2 + \|D_\mu D_\theta U_h^*(t_j)\|^2\Big] \\ &\le C\big[\|x\|^2 + \|\sigma(\cdot)\|^2_{C([0,T];\mathbb{L}^2)}\big]\,, \end{aligned}$$

which leads to $\tau\sum_{j=0}^{n-1} A_0^{n-j}\prod_{k=j+2}^{n}(1+\Delta_k W)\, U_h^*(t_j) \in \mathbb{D}^{2,2}(\mathbb{L}^2)$. Hence, for any $t\in[0,T]$, $\mathcal{S}_{h\tau}\big(v(t);\Pi_\tau U_h^*\big) \in \mathbb{D}^{2,2}(\mathbb{L}^2)$, and

$$
\begin{aligned}
&\sup_{\theta\in[0,T]}\sup_{\nu(t)\in[\theta,T]}\mathbb{E}\Big[\big\|D_\theta\mathcal{S}_{h\tau}\big(\nu(t);\,\Pi_\tau U_h^*\big)\big\|^2\Big]\\
&\quad+\sup_{\mu,\theta\in[0,T]}\sup_{\nu(t)\in[\mu\vee\theta,T]}\mathbb{E}\Big[\big\|D_\mu D_\theta\mathcal{S}_{h\tau}\big(\nu(t);\,\Pi_\tau U_h^*\big)\big\|^2\Big]\\
&\qquad\le C\Big[\|x\|^2+\|\sigma(\cdot)\|^2_{C([0,T];\mathbb{L}^2)}\Big]. \qquad (3.87)
\end{aligned}
$$

Applying the same trick, the optimality condition (3.36), (3.61a) and (3.62b) with $\gamma = 1$ in Lemma 3.8 we can deduce that there exists a constant C such that for any $\theta \in [0, T]$,

$$
\mathbb{E}\big[\|\nabla D_\theta\mathcal{S}_{h\tau}(T;\,\Pi_\tau U_h^*)\|^2\big]\le C\big[\|x\|^2_{\mathbb{H}_0^1}+\|\sigma(\cdot)\|^2_{C([0,T];\mathbb{H}_0^1)}\big]. \qquad (3.88)
$$

(2) By the same procedure as in the proof of (3.62a), (3.62b) in Lemma 3.8, thanks to (3.87), we can obtain

$$
\begin{aligned}
&\sup_{\theta\in[0,T]}\mathbb{E}\Big[\int_\theta^T\|D_\theta Z_h(t;\,\mathcal{S}_{h\tau}(\Pi_\tau U_h^*))\|^2\,\mathrm{d}t\Big]\\
&\quad\le C\sup_{\theta\in[0,T]}\sup_{\nu(t)\in[\theta,T]}\mathbb{E}\big[\|D_\theta\mathcal{S}_{h\tau}(\nu(t);\,\Pi_\tau U_h^*)\|^2\big]\\
&\quad\le C\Big[\|x\|^2+\|\sigma(\cdot)\|^2_{C([0,T];\mathbb{L}^2)}\Big], \qquad (3.89)
\end{aligned}
$$

$$
\begin{aligned}
&\sup_{\mu,\theta\in[0,T]}\sup_{t\in[\mu\vee\theta,T]}\mathbb{E}\Big[\big\|D_\mu D_\theta Y_h\big(t;\,\mathcal{S}_{h\tau}(\Pi_\tau U_h^*)\big)\big\|^2\Big]\\
&\quad\le C\sup_{\mu,\theta\in[0,T]}\sup_{\nu(t)\in[\mu\vee\theta,T]}\mathbb{E}\Big[\big\|D_\mu D_\theta\mathcal{S}_{h\tau}\big(\nu(t);\,\Pi_\tau U_h^*\big)\big\|^2\Big]\\
&\quad\le C\Big[\|x\|^2+\|\sigma(\cdot)\|^2_{C([0,T];\mathbb{L}^2)}\Big], \qquad (3.90)
\end{aligned}
$$

and

$$
\begin{aligned}
&\sup_{\theta\in[0,T]}\mathbb{E}\Big[\int_\theta^T\big\|\Delta_h D_\theta Y_h\big(t;\,\mathcal{S}_{h\tau}(\Pi_\tau U_h^*)\big)\big\|^2\,\mathrm{d}t\Big]\\
&\quad\le C\sup_{\theta\in[0,T]}\sup_{\nu(t)\in[\theta,T]}\mathbb{E}\Big[\big\|\nabla D_\theta\mathcal{S}_{h\tau}\big(\nu(t);\,\Pi_\tau U_h^*\big)\big\|^2\Big]\\
&\quad\le C\big[\|x\|^2_{\mathbb{H}_0^1}+\|\sigma(\cdot)\|^2_{C([0,T];\mathbb{H}_0^1)}\big]. \qquad (3.91)
\end{aligned}
$$

(3) Applying the fact that $Z_h\big(\cdot;\,\mathcal{S}_{h\tau}(\Pi_\tau U_h^*)\big) = D_\cdot Y_h\big(\cdot;\,\mathcal{S}_{h\tau}(\Pi_\tau U_h^*)\big)$ a.e., for any $t\in[t_n,t_{n+1})$, $n = 0, 1, \cdots, N-1$, we arrive at

$$
\begin{aligned}
&\mathbb{E}\Big[\big\|Z_h\big(t;\,\mathcal{S}_{h\tau}(\Pi_\tau U_h^*)\big)-Z_h\big(\nu(t);\,\mathcal{S}_{h\tau}(\Pi_\tau U_h^*)\big)\big\|^2\Big]\\
&\quad\le 2\mathbb{E}\Big[\big\|(D_t-D_{t_n})Y_h\big(t;\,\mathcal{S}_{h\tau}(\Pi_\tau U_h^*)\big)\big\|^2
\end{aligned}
$$

$$+\left\|D_{t_n}\Big(Y_h\big(t;\mathcal{S}_{h\tau}(\Pi_\tau U_h^*)\big)-Y_h\big(t_n;\mathcal{S}_{h\tau}(\Pi_\tau U_h^*)\big)\Big)\right\|^2\Big]$$
$$=: 2(I_1+I_2)\,. \tag{3.92}$$

Similar to (3.62c) in Lemma 3.8,

$$I_1\le C\max_{n\le i\le N}\mathbb{E}\big[\|(D_t-D_{t_n})\mathcal{S}_{h\tau}(t_i;\Pi_\tau U_h^*)\|^2\big]\,,$$

which, together with

$$\begin{aligned}&(D_t-D_{t_n})\mathcal{S}_{h\tau}(t_i;\Pi_\tau U_h^*)\\&=\tau\sum_{j=0}^{i-1}A_0^{i-j}\prod_{k=j+2}^{i}(1+\beta\Delta_k W)\,(D_t-D_{t_n})U_h^*(t_j)\quad\text{a.e.}\,,\end{aligned}$$

the optimality condition (3.36), and (3.62c) in Lemma 3.8, yields

$$\begin{aligned}I_1&\le C\max_{n\le j\le N}\mathbb{E}\big[\|(D_t-D_{t_n})U_h^*(t_j)\|^2\big]\\&\le C|t-t_n|\big[\|x\|^2_{\mathbb{H}_0^1}+\|\sigma(\cdot)\|^2_{C([0,T];\mathbb{H}_0^1)}+L^2_{\sigma,1/2}\big]\,.\end{aligned}\tag{3.93}$$

By virtue of (3.87), (3.89)–(3.91), we can get

$$\begin{aligned}I_2\le C\Big\{&\mathbb{E}\Big[\int_{t_n}^{t}\big\|D_{t_n}D_\theta Y_h\big(\theta;\mathcal{S}_{h\tau}(\Pi_\tau U_h^*)\big)\big\|^2\,\mathrm{d}\theta\Big]\\&+(t-t_n)\mathbb{E}\Big[\int_{t_n}^{T}\big\|\Delta_h D_{t_n}Y_h\big(\theta;\mathcal{S}_{h\tau}(\Pi_\tau U_h^*)\big)\big\|^2+\big\|D_{t_n}Z_h\big(\theta;\mathcal{S}_{h\tau}(\Pi_\tau U_h^*)\big)\big\|^2\\&\qquad+\big\|D_{t_n}\mathcal{S}_{h\tau}\big(\nu(\theta);\Pi_\tau U_h^*\big)\big\|^2\,\mathrm{d}\theta\Big]\Big\}\\\le{}& C|t-t_n|\big[\|x\|^2_{\mathbb{H}_0^1}+\|\sigma(\cdot)\|^2_{C([0,T];\mathbb{H}_0^1)}\big]\,.\end{aligned}\tag{3.94}$$

Now the desired result (3.86) can be derived by (3.92)–(3.94). □

Lemma 3.13 *Suppose that* $\big(Y_h\big(\cdot;\mathcal{S}_{h\tau}(\Pi_\tau U_h^*)\big), Z_h\big(\cdot;\mathcal{S}_{h\tau}(\Pi_\tau U_h^*)\big)\big)$ *solves* (3.37) *with* $X_h^*(\cdot)=\mathcal{S}_{h\tau}(\cdot;\Pi_\tau U_h^*)$, *and* $\big(Y_\cdot\big(\mathcal{S}_{h\tau}(\cdot;\Pi_\tau U_h^*)\big), Z_0\big(\cdot;\mathcal{S}_{h\tau}(\cdot;\Pi_\tau U_h^*)\big)\big)$ *solves* (3.80) *with* $X_{h\tau}^*(\cdot)=\mathcal{S}_{h\tau}(\cdot;\Pi_\tau U_h^*)$. *Then, there exist a constant* C *independent of* h , τ *such that*

$$\begin{aligned}&\sum_{n=0}^{N-1}\mathbb{E}\Big[\int_{t_n}^{t_{n+1}}\big\|Y_h\big(t;\mathcal{S}_{h\tau}(\Pi_\tau U_h^*)\big)-Y_n\big(\mathcal{S}_{h\tau}(\Pi_\tau U_h^*)\big)\big\|^2\,\mathrm{d}t\Big]\\&\qquad\le C\tau\big[\|x\|^2_{\mathbb{H}_0^1\cap\mathbb{H}^2}+\|\sigma(\cdot)\|^2_{C([0,T];\mathbb{H}_0^1)\cap L^2(0,T;\mathbb{H}_0^1\cap\mathbb{H}^2)}+L^2_{\sigma,1/2}\big]\,.\end{aligned}$$

Proof For simplicity, in the proof we write $\big(Y_h\big(\cdot; \mathcal{S}_{h\tau}(\Pi_\tau U_h^*)\big), Z_h\big(\cdot; \mathcal{S}_{h\tau}(\Pi_\tau U_h^*)\big)\big)$ resp. $\big(Y_\cdot\big(\mathcal{S}_{h\tau}(\Pi_\tau U_h^*)\big), Z_0\big(\cdot; \mathcal{S}_{h\tau}(\Pi_\tau U_h^*)\big)\big)$ as $\big(\widetilde{Y}_h(\cdot), \widetilde{Z}_h(\cdot)\big)$ resp. $\big(\widetilde{Y}_n, \widetilde{Z}_0(\cdot)\big)$.

Firstly we apply Theorem 3.5 with $Y_{T,h} = \widetilde{Y}_{Y,h} = -\alpha\mathcal{S}_{h\tau}(T; \Pi_\tau U_h^*)$ and $f_h(\cdot) = \widetilde{f}_h(\cdot) = \mathcal{S}_{h\tau}(\cdot; \Pi_\tau U_h^*)$, and arrive at

$$
\begin{aligned}
&\sum_{n=0}^{N-1} \mathbb{E}\Big[\int_{t_n}^{t_{n+1}} \|\widetilde{Y}_h(t) - \widetilde{Y}_n\|^2\, \mathrm{d}t\Big] \\
&\leq C\tau\mathbb{E}\Big[\|\mathcal{S}_{h\tau}(T; \Pi_\tau U_h^*)\|^2_{\mathbb{H}_h^2} + \int_0^T \|\mathcal{S}_{h\tau}(t; \Pi_\tau U_h^*)\|^2_{\dot{\mathbb{H}}_h^1}\, \mathrm{d}t\Big] \\
&\quad + C\sum_{n=0}^{N-1} \mathbb{E}\Big[\int_{t_n}^{t_{n+1}} \|\widetilde{Z}_h(t) - \overline{\widetilde{Z}}_h(t)\|^2_{\dot{\mathbb{H}}_h^{-1}}\, \mathrm{d}t\Big] \\
&=: C\tau\sum_{i=1}^{2} I_i + CI_3\,.
\end{aligned}
$$

For I_1, by Theorem 3.7, Lemmata 2.2 and 3.9, we find that

$$
I_1 \leq C\big[\|x\|^2_{\mathbb{H}_0^1\cap\mathbb{H}^2} + \|\sigma(\cdot)\|^2_{L^2(0,T;\mathbb{H}_0^1\cap\mathbb{H}^2)}\big]\,.
$$

Term I_2 can be estimated in the same vein and

$$
I_2 \leq C\big[\|x\|^2_{\mathbb{H}_0^1} + \|\sigma(\cdot)\|^2_{L^2(0,T;\mathbb{H}_0^1)}\big]\,.
$$

For I_3, by Lemma 3.12 we have

$$
\begin{aligned}
I_3 &\leq \sum_{n=0}^{N-1} \mathbb{E}\Big[\int_{t_n}^{t_{n+1}} \|\widetilde{Z}_h(t) - \widetilde{Z}_h(t_n)\|^2_{\dot{\mathbb{H}}_h^{-1}}\, \mathrm{d}t\Big] \\
&\leq C\tau\big[\|x\|^2_{\mathbb{H}_0^1} + \|\sigma(\cdot)\|^2_{C([0,T];\mathbb{H}_0^1)} + L^2_{\sigma,1/2}\big]\,.
\end{aligned}
$$

A combination of estimates for I_1 through I_3 settles the desired assertion. □

We are now ready to verify rates of convergence for the optimal pair of Problem **(SLQ)**$_{h\tau}$; it is as in Sect. 3.3.1 that $\mathcal{S}_{h\tau} : \mathbb{U}_\mathbb{F} \to \mathbb{X}_\mathbb{F}$ is used, which is the solution operator to the forward equation (3.70a).

Proof of Theorem 3.8. We divide the proof into two steps.

(1) We follow the argument in the proof of Theorem 3.6. Firstly, we have

$$
\begin{aligned}
&\|U_{h\tau}^*(\cdot) - \Pi_\tau U_h^*(\cdot)\|^2_{L^2_\mathbb{F}(0,T;\mathbb{L}^2)} \\
&\quad\leq \Big[\Big(D\mathcal{J}_h\big(U_h^*(\cdot)\big) - D\mathcal{J}_h\big(\Pi_\tau U_h^*(\cdot)\big), U_{h\tau}^*(\cdot) - \Pi_\tau U_h^*(\cdot)\Big)_{L^2_\mathbb{F}(0,T;\mathbb{L}^2)} \\
&\qquad + \Big(D\mathcal{J}_h\big(\Pi_\tau U_h^*(\cdot)\big) - D\mathcal{J}_{h,\tau}\big(\Pi_\tau U_h^*(\cdot)\big), U_{h\tau}^*(\cdot) - \Pi_\tau U_h^*(\cdot)\Big)_{L^2_\mathbb{F}(0,T;\mathbb{L}^2)}\Big]\,.
\end{aligned} \tag{3.95}
$$

Therefore,

$$
\begin{aligned}
&\|U_{h\tau}^* - \Pi_\tau U_h^*\|_{L^2_{\mathbb{F}}(0,T;\mathbb{L}^2)}^2 \\
&\quad \le 3\Big[\big\|D\mathcal{J}_h\big(U_h^*(\cdot)\big) - D\mathcal{J}_h\big(\Pi_\tau U_h^*(\cdot)\big)\big\|_{L^2_{\mathbb{F}}(0,T;\mathbb{L}^2)}^2 \\
&\qquad + \big\|\mathcal{T}_h^1\big(\cdot;\mathcal{S}_h(\Pi_\tau U_h^*)\big) - Y_\cdot\big(\mathcal{S}_{h\tau}(\Pi_\tau U_h^*)\big)\big\|_{L^2_{\mathbb{F}}(0,T;\mathbb{L}^2)}^2 \\
&\qquad + \big\|Y_\cdot\big(\mathcal{S}_{h\tau}(\Pi_\tau U_h^*)\big) - Y_{h\tau}\big(\cdot;\mathcal{S}_{h\tau}(\Pi_\tau U_h^*)\big)\big\|_{L^2_{\mathbb{F}}(0,T;\mathbb{L}^2)}^2\Big] \\
&\quad = 3\sum_{i=1}^{3} I_i\,.
\end{aligned} \tag{3.96}
$$

We use (3.41) and the optimality condition (3.36) to bound I_1 as follows,

$$
\begin{aligned}
I_1 \le 2\big[&\|U_h^*(\cdot) - \Pi_\tau U_h^*(\cdot)\|_{L^2_{\mathbb{F}}(0,T;\mathbb{L}^2)}^2 \\
&+ \big\|\mathcal{T}_h^1\big(\cdot;\mathcal{S}_h(\Pi_\tau U_h^*)\big) - \mathcal{T}_h^1\big(\cdot;\mathcal{S}_h(U_h^*)\big)\big\|_{L^2_{\mathbb{F}}(0,T;\mathbb{L}^2)}^2\big]\,.
\end{aligned} \tag{3.97}
$$

By Lemma 3.3 on stability properties of solutions to BSDE (3.37) with $X_h^*(\cdot) = \mathcal{S}_h(\cdot;\,\Pi_\tau U_h^*)$ and Lemma 3.1 for SDE (3.1) with $x = 0$, $f_h(\cdot) = U_h^*(\cdot) - \Pi_\tau U_h^*(\cdot)$, $g_h(\cdot) = 0$, we obtain

$$
\begin{aligned}
&\big\|\mathcal{T}_h^1\big(\cdot;\mathcal{S}_h(\Pi_\tau U_h^*)\big) - \mathcal{T}_h^1\big(\cdot;\mathcal{S}_h(U_h^*)\big)\big\|_{L^2_{\mathbb{F}}(0,T;\mathbb{L}^2)}^2 \\
&\quad \le C\Big[\|\mathcal{S}_h(T;\,U_h^*) - \mathcal{S}_h(T;\,\Pi_\tau U_h^*)\|_{L^2_{\mathcal{F}_T}(\Omega;\dot{\mathbb{H}}_h^{-1})}^2 \\
&\qquad + \|\mathcal{S}_h(\cdot;\,U_h^*) - \mathcal{S}_h(\cdot;\,\Pi_\tau U_h^*)\|_{L^2_{\mathbb{F}}(0,T;\dot{\mathbb{H}}_h^{-2})}^2\Big] \\
&\quad \le C\|U_h^*(\cdot) - \Pi_\tau U_h^*(\cdot)\|_{L^2_{\mathbb{F}}(0,T;\mathbb{L}^2)}^2\,.
\end{aligned} \tag{3.98}
$$

By the optimality condition (3.36), applying estimate (3.15) of Lemma 3.3 with $Y_{T,h} = -\alpha\mathcal{S}_h(T;\,U_h^*)$, $f_h(\cdot) = \mathcal{S}_h(\cdot;\,U_h^*)$, then Lemma 3.1 with $f_h(\cdot) = U_h^*(\cdot)$, $g_h(\cdot) = \Pi_h\sigma(\cdot)$, and (3.56b) in Lemma 3.7, we have

$$
\begin{aligned}
&\|U_h^*(\cdot) - \Pi_\tau U_h^*(\cdot)\|_{L^2_{\mathbb{F}}(0,T;\mathbb{L}^2)}^2 = \|Y_h(\cdot) - \Pi_\tau Y_h(\cdot)\|_{L^2_{\mathbb{F}}((0,T;\mathbb{L}^2)}^2 \\
&\quad \le C\tau\mathbb{E}\Big[\|\mathcal{S}_h(T;\,U_h^*)\|_{\mathbb{H}_0^1}^2 + \int_0^T \|\mathcal{S}_h(t;\,U_h^*)\|^2\,\mathrm{d}t\Big] \\
&\quad \le C\tau\big[\|x\|_{\mathbb{H}_0^1}^2 + \|\sigma(\cdot)\|_{L^2(0,T;\mathbb{H}_0^1)}^2\big]\,.
\end{aligned} \tag{3.99}
$$

Next, we turn to I_2. The triangle inequality leads to

$$
\begin{aligned}
I_2 \le 2\Big[&\big\|\mathcal{T}_h^1\big(\cdot;\mathcal{S}_h(\Pi_\tau U_h^*)\big) - \mathcal{T}_h^1\big(\cdot;\mathcal{S}_{h\tau}(\Pi_\tau U_h^*)\big)\big\|_{L^2_{\mathbb{F}}(0,T;\mathbb{L}^2)}^2 \\
&+\big\|\mathcal{T}_h^1\big(\cdot;\mathcal{S}_{h\tau}(\Pi_\tau U_h^*)\big) - Y_\cdot\big(\mathcal{S}_{h\tau}(\Pi_\tau U_h^*)\big)\big\|_{L^2_{\mathbb{F}}(0,T;\mathbb{L}^2)}^2\Big] \\
=:\ & 2\big(I_{21} + I_{22}\big)\,.
\end{aligned}
$$

In order to bound I_{21}, we use Lemma 3.3, then (3.10) in Theorem 3.2 with $f_h(\cdot) = \widetilde{f}_h(\cdot) = \Pi_\tau U_h^*(\cdot)$, $g_h(\cdot) = \widetilde{g}_h(\cdot) = \Pi_h \sigma(\cdot)$, and finally (3.61a) in Lemma 3.8 to conclude that

$$\begin{aligned} I_{21} &\leq C\Big\{ \mathbb{E}\big[\|\mathcal{S}_h(T; \Pi_\tau U_h^*) - \mathcal{S}_{h\tau}(T; \Pi_\tau U_h^*)\|^2 \big] \\ &\quad + \mathbb{E}\Big[\int_0^T \|\mathcal{S}_h(t; \Pi_\tau U_h^*) - \mathcal{S}_{h\tau}(t; \Pi_\tau U_h^*)\|^2_{\dot{\mathbb{H}}_h^{-1}} \,\mathrm{d}t \Big] \Big\} \\ &\leq C\tau \Big[\|x\|^2_{\mathbb{H}_0^1 \cap \mathbb{H}^2} + \|\sigma(\cdot)\|^2_{L^2(0,T;\mathbb{H}_0^1 \cap \mathbb{H}^2)} \Big]. \end{aligned}$$

For I_{22}, Lemma 3.13 implies that

$$I_{22} \leq C\tau \big[\|x\|^2_{\mathbb{H}_0^1 \cap \mathbb{H}^2} + \|\sigma(\cdot)\|^2_{C([0,T];\mathbb{H}_0^1) \cap L^2(0,T;\mathbb{H}_0^1 \cap \mathbb{H}^2)} + L^2_{\sigma,1/2} \big].$$

Finally, Lemma 3.11 leads to

$$I_3 \leq C\tau \big[\|x\|^2 + \|\sigma(\cdot)\|^2_{C([0,T];\mathbb{L}^2)} \big].$$

Now we insert above estimates into (3.96) to obtain the $U_h^*(\cdot) - U_{h\tau}^*(\cdot)$ part of the assertion (3.72).

(2) By setting $f_h(\cdot) = U_h^*(\cdot)$, $\widetilde{f}_h(\cdot) = U_{h\tau}^*(\cdot)$ and $g_h(\cdot) = \Pi_h \sigma(\cdot)$, $\widetilde{g}_h(\cdot) = \Pi_\tau \Pi_h \sigma(\cdot)$, and applying the assertion (3.72) as well as Theorem 3.2, (3.56b) in Lemma 3.7, we deduce that

$$\begin{aligned} &\sum_{n=0}^{N-1} \mathbb{E}\Big[\int_{t_n}^{t_{n+1}} \|X_h^*(t) - X_{h\tau}^*(t_n)\|^2 \,\mathrm{d}t \Big] \\ &\leq C\tau \Big\{ \|x\|^2_{\mathbb{H}_0^1} + \mathbb{E}\Big[\int_0^T \|U_h^*(t)\|^2 + \|\sigma(t)\|^2_{\mathbb{H}_0^1} \,\mathrm{d}t \Big] \Big\} \\ &\quad + C \sum_{n=0}^{N-1} \mathbb{E}\Big[\int_{t_n}^{t_{n+1}} \|U_h^*(t) - U_{h\tau}^*(t_n)\|^2 + \|\sigma(t) - \sigma(t_n)\|^2 \,\mathrm{d}t \Big] \\ &\leq C\tau \big[\|x\|^2_{\mathbb{H}_0^1 \cap \mathbb{H}^2} + \|\sigma(\cdot)\|^2_{C([0,T];\mathbb{H}_0^1) \cap L^2(0,T;\mathbb{H}_0^1 \cap \mathbb{H}^2)} + L^2_{\sigma,1/2} \big], \end{aligned}$$

which is the $X_h^*(\cdot) - X_{h\tau}^*(\cdot)$ part of the assertion 3.72. Moreover, applying 3.3 with $\gamma = 1$ in Lemma 3.1, together with 2.4d in Lemma 2.2, we further obtain

$$\begin{aligned} &\max_{0 \leq n \leq N} \mathbb{E}\big[\|X_h^*(t_n) - X_{h\tau}^*(t_n)\|^2 \big] + \tau \sum_{n=0}^{N-1} \mathbb{E}\big[\|X_h^*(t_n) - X_{h\tau}^*(t_n)\|^2_{\mathbb{H}_0^1} \big] \\ &\leq C\tau \Big\{ \|x\|^2_{\mathbb{H}_0^1 \cap \mathbb{H}^2} + \mathbb{E}\Big[\int_0^T \|U_h^*(t)\|^2_{\mathbb{H}_0^1} + \|\sigma(t)\|^2_{\mathbb{H}_0^1 \cap \mathbb{H}^2} \,\mathrm{d}t \Big] \Big\} \end{aligned}$$

$$+C\sum_{n=0}^{N-1}\mathbb{E}\Big[\int_{t_n}^{t_{n+1}}\|U_h^*(t)-U_{h\tau}^*(t_n)\|^2+\|\sigma(t)-\sigma(t_n)\|^2\,\mathrm{d}t\Big]$$

$$\leq C\tau\Big[\|x\|^2_{\mathbb{H}_0^1\cap\mathbb{H}^2}+\|\sigma(\cdot)\|^2_{C([0,T];\mathbb{H}_0^1)\cap L^2(0,T;\mathbb{H}_0^1\cap\mathbb{H}^2)}+L^2_{\sigma,1/2}\Big].$$

That settles the assertion (3.73) and completes the proof.

3.3.4 *The Gradient Descent Method for Problem* $(\mathbf{SLQ})_{h\tau}$

By Theorem 3.7, solving Problem $(\mathbf{SLQ})_{h\tau}$ is equivalent to solving the system of coupled forward-backward difference equations (3.70) and (3.71). We may exploit the variational character of Problem $(\mathbf{SLQ})_{h\tau}$ to construct a gradient descent method where approximate iterates of the optimal control $U_{h\tau}^*(\cdot)$ in the Hilbert space $\mathbb{U}_{\mathbb{F}}$ are obtained; see also [18, 19] for more details.

Algorithm 3.9 *Let* $U_{h\tau}^{(0)}(\cdot)\in\mathbb{U}_{\mathbb{F}}$, *and fix* $\kappa>0$. *For any* $\ell\in\mathbb{N}_0$, *update* $U_{h\tau}^{(\ell)}(\cdot)\in\mathbb{U}_{\mathbb{F}}$ *as follows:*

1. *Compute* $X_{h\tau}^{(\ell)}(\cdot)\in\mathbb{X}_{\mathbb{F}}$ *by*

$$\begin{cases}[\mathbb{1}-\tau\Delta_h]X_{h\tau}^{(\ell)}(t_{n+1})=X_{h\tau}^{(\ell)}(t_n)+\tau U_{h\tau}^{(\ell)}(t_n)\\ \qquad+\big[\beta X_{h\tau}^{(\ell)}(t_n)+\Pi_h\sigma(t_n)\big]\Delta_{n+1}W\quad n=0,1,\cdots,N-1,\\ X_{h\tau}^{(\ell)}(0)=\Pi_h x\,.\end{cases}\tag{3.100}$$

2. *Use* $X_{h\tau}^{(\ell)}(\cdot)\in\mathbb{X}_{\mathbb{F}}$ *to compute* $Y_{h\tau}^{(\ell)}(\cdot)\in\mathbb{X}_{\mathbb{F}}$ *via*

$$\begin{aligned}Y_{h\tau}^{(\ell)}(t_n)=&-\tau\mathbb{E}^{t_n}\Big[\sum_{j=n+1}^{N-1}A_0^{j-n}\prod_{k=n+2}^{j}(1+\beta\Delta_k W)X_{h\tau}^{(\ell)}(t_j)\Big]\\ &-\alpha\mathbb{E}^{t_n}\Big[A_0^{N-n}\prod_{k=n+2}^{N}(1+\beta\Delta_k W)X_{h\tau}^{(\ell)}(T)\Big]\quad n=0,1,\cdots,N-1\,.\end{aligned}$$

3. *Update* $U_{h\tau}^{(\ell+1)}(\cdot)\in\mathbb{U}_{\mathbb{F}}$ *via*

$$U_{h\tau}^{(\ell+1)}(\cdot)=U_{h\tau}^{(\ell)}(\cdot)-\frac{1}{\kappa}\big[U_{h\tau}^{(\ell)}(\cdot)-Y_{h\tau}^{(\ell)}(\cdot)\big].$$

If compared with (3.70)–(3.71), steps 1 and 2 are now decoupled: the first step requires to solve a spatio-temporal discretized equation of SPDE (2.18a), while the second requires to solve a spatio-temporal discretized equation of the BSPDE (2.18b) which is not the numerical solution by the implicit Euler method (see Lemmata 3.10

and 3.11 for the difference). A similar method to solve Problem **(SLQ)**$_{h\tau}$ has been proposed in [5, 11].

Below, we present a lower bound for κ and discuss the speed of convergence of Algorithm 3.9. For this purpose, we first show the Lipschitz continuity of $D\mathcal{J}_{h,\tau}(\cdot)$. By borrowing notations defined in (3.75)–(3.76) in the proof of Theorem 3.7, and applying (3.77) and (3.79), we know that

$$\begin{aligned} D\mathcal{J}_{h,\tau}\big(U_{h\tau}(\cdot)\big) &= U_{h\tau}(\cdot) + L^*\big[\big(\Gamma X_{h\tau}(0)\big)(\cdot) + \big(LU_{h\tau}(\cdot)\big)(\cdot) + f(\cdot)\big](\cdot) \\ &\quad + \alpha\widehat{L}^*[\widehat{\Gamma} X_{h\tau}(0) + \widehat{L}U_{h\tau} + \widehat{f}\,](\cdot)\,. \end{aligned}$$

Thus by the linearity of $L\,,\widehat{L}\,,\Gamma\,,\widehat{\Gamma}$, for any $U^1_{h\tau}(\cdot)\,,U^2_{h\tau}(\cdot)\in\mathbb{U}_{\mathbb{F}}$, it follows that $\big\|D\mathcal{J}_{h,\tau}\big(U^1_{h\tau}(\cdot)\big) - D\mathcal{J}_{h,\tau}\big(U^2_{h\tau}(\cdot)\big)\big\|_{\mathbb{U}_{\mathbb{F}}} \le \|\mathbb{1}_h + L^*L + \alpha\widehat{L}^*\widehat{L}\|_{\mathcal{L}(\mathbb{U}_{\mathbb{F}})}\|U^1_{h\tau}(\cdot) - U^2_{h\tau}(\cdot)\|_{\mathbb{U}_{\mathbb{F}}}$. Furthermore, we can show a upper bound of $\|\mathbb{1}_h + L^*L + \alpha\widehat{L}^*\widehat{L}\|_{\mathcal{L}(\mathbb{U}_{\mathbb{F}})}$. Indeed, on noting that $\|A_0\|_{\mathcal{L}(\mathbb{V}_h|_{\mathbb{L}^2})} = \|(\mathbb{1}_h - \tau\Delta_h)^{-1}\|_{\mathcal{L}(\mathbb{V}_h|_{\mathbb{L}^2})} \le 1$, we find that

$$\begin{aligned} \big\|\big(LU_{h\tau}(\cdot)\big)(\cdot)\big\|^2_{\mathbb{X}_{\mathbb{F}}} &= \sum_{n=0}^{N-1}\tau\mathbb{E}\Big[\Big\|\tau\sum_{j=0}^{n-1}A_0^{n-j}\prod_{k=j+2}^{n}(1+\beta\Delta_k W)\,U_{h\tau}(t_j)\Big\|^2\Big] \\ &\le T^2e^{\beta^2T}\|U_{h\tau}(\cdot)\|^2_{\mathbb{U}_{\mathbb{F}}}\,. \end{aligned}$$

In a similar vein, we can prove that $\|\widehat{L}U_{h\tau}(\cdot)\|^2_{L^2_{\mathcal{F}_T}(\Omega;\mathbb{L}^2)} \le Te^{\beta^2T}\|U_{h\tau}\|^2_{\mathbb{U}_{\mathbb{F}}}$. Hence

$$\|\mathbb{1}_h + L^*L + \alpha\widehat{L}^*\widehat{L}\|_{\mathcal{L}(\mathbb{U}_{\mathbb{F}})} \le 1 + \alpha Te^{\beta^2T} + T^2e^{\beta^2T}\,.$$

Since Algorithm 3.9 is the gradient descent method for Problem **(SLQ)**$_{h\tau}$, we have the following estimates.

Theorem 3.10 *Suppose that* $\kappa \ge \|\mathbb{1}_h + L^*L + \alpha\widehat{L}^*\widehat{L}\|_{\mathcal{L}(\mathbb{U}_{\mathbb{F}})}$. *Let* $\{U^{(\ell)}_{h\tau}(\cdot)\}_{\ell\in\mathbb{N}_0}\subset\mathbb{U}_{\mathbb{F}}$ *be generated by Algorithm 3.9, and* $U^*_{h\tau}(\cdot)$ *solves Problem* **(SLQ)**$_{h\tau}$. *Then, for* $\ell = 1, 2, \cdots$,

$$\begin{cases} \big\|U^{(\ell)}_{h\tau}(\cdot) - U^*_{h\tau}(\cdot)\big\|^2_{\mathbb{U}_{\mathbb{F}}} \le \Big(1-\dfrac{1}{\kappa}\Big)^{\ell}\big\|U^{(0)}_{h\tau}(\cdot) - U^*_{h\tau}(\cdot)\big\|^2_{\mathbb{U}_{\mathbb{F}}}\,, & (3.101\text{a}) \\[2ex] \mathcal{J}_{h,\tau}\big(U^{(\ell)}_{h\tau}(\cdot)\big) - \mathcal{J}_{h,\tau}\big(U^*_{h\tau}(\cdot)\big) \le \dfrac{2\kappa\big\|U^{(0)}_{h\tau}(\cdot) - U^*_{h\tau}(\cdot)\big\|^2_{\mathbb{U}_{\mathbb{F}}}}{\ell}\,, & (3.101\text{b}) \\[2ex] \max\limits_{0\le n\le N}\mathbb{E}\big[\big\|X^*_{h\tau}(t_n) - X^{(\ell)}_{h\tau}(t_n)\big\|^2\big] + \tau\sum\limits_{n=0}^{N-1}\mathbb{E}\big[\big\|X^*_{h\tau}(t_n) - X^{(\ell)}_{h\tau}(t_n)\big\|^2_{\mathbb{H}^1_0}\big] & \\ \qquad \le C\Big(1-\dfrac{1}{\kappa}\Big)^{\ell}\big\|U^{(0)}_{h\tau}(\cdot) - U^*_{h\tau}(\cdot)\big\|^2_{\mathbb{U}_{\mathbb{F}}}\,, & (3.101\text{c}) \end{cases}$$

where C *is independent of* $h,\ \tau,\ \ell$.

Proof The estimates (3.101a) and (3.101b) are standard for the gradient descent method (see e.g. [18, Theorem 1.2.4]); more details can also be found in [11, Sect. 5]. In the following, we restrict to assertion (3.101c).

For all $n = 0, 1, \cdots, N$, define $\bar{e}_X^{n,\ell} = X_{h\tau}^*(t_n) - X_{h\tau}^{(\ell)}(t_n)$. Subtracting (3.100) from (3.69) (where $U_{h\tau}(\cdot) = U_{h\tau}^*(\cdot)$) leads to

$$\bar{e}_X^{n+1,\ell} - \bar{e}_X^{n,\ell} = \tau\Delta_h\bar{e}_X^{n+1,\ell} + \beta\bar{e}_X^{n,\ell}\Delta_{n+1}W + \tau\big[U_{h\tau}^*(t_n) - U_{h\tau}^{(\ell)}(t_n)\big].$$

Following the procedure as that in the proof of Lemma 3.2, we can derive (3.101c). □

3.4 Implementable Modifications of the Gradient Descent Method

Algorithm 3.9 is an iterative method to approximate the optimal control $U_{h\tau}^*(\cdot)$ of Problem $\textbf{(SLQ)}_{h\tau}$—and its benefit lies in the decoupled computation of iterates which approximate the optimal state, control, and (adjoint) states. However, it is still not practical, since step 2 of Algorithm 3.9 involves *conditional expectations* $\mathbb{E}^{t_n}[\cdot]$.

In this section, we present two modifications of Algorithm 3.9 which are implementable: the first uses an *exact* representation of the conditional expectation in its step 2; as a consequence, the rates obtained in Sect. 3.3 remain valid here. As will be detailed in Theorem 3.11 of Sect. 3.4.1, this strategy is possible for $\beta = 0$. The second strategy is proposed in Sect. 3.4.2: it uses an approximation of each $\mathbb{E}^{t_n}[\cdot]$ by statistical regression, and therefore introduces a further (statistical) error, whose estimation will be left open. In comparative simulations, it turns out that the second method requires far more computational resources, but is applicable for general $\beta \in \mathbb{R}$.

3.4.1 Exact Representation of Conditional Expectations

Let $\beta = 0$ in Problem $\textbf{(SLQ)}_{h\tau}$. The following theorem asserts that the updated control iterate $U_{h\tau}^{(\ell+1)}(\cdot)$ in step 3 of Algorithm 3.9 may be found *without* the computation of conditional expectations $\mathbb{E}^{t_n}[\cdot]$—as is stated in step 2, where the new adjoint $Y_{h\tau}^{(\ell)}(\cdot)$ is determined. This idea was first exploited in the context of stochastic boundary control [20]. In fact, $Y_{h\tau}^{(\ell)}(\cdot)$ now satisfies the identity (3.102) below as well, which does not involve conditional expectations.

Theorem 3.11 *Suppose that $\beta = 0$. In the Algorithm 3.9, if $U_{h\tau}^{(0)}(\cdot)$ is of the form*

$$U_{h\tau}^{(0)}(t_n) = \sum_{m=1}^{n}\prod_{i=1}^{m}(1+\Delta_i W) f_{n,m}^{(0)} + \sum_{m=1}^{n}\Delta_m W \widetilde{f}_{n,m}^{(0)} + g_n^{(0)} \quad n=0,1,\cdots,N-1,$$

where $f_{n,m}^{(0)}$, $\widetilde{f}_{n,m}^{(0)}$, $g_n^{(0)}$ are deterministic for $m=1,2,\cdots,n$. Then for any $\ell \in \mathbb{N}_0$, it holds that

$$Y_{h\tau}^{(\ell)}(t_n) = \sum_{m=1}^{n}\prod_{i=1}^{m}(1+\Delta_i W) F_{n,m}^{(\ell)} + \sum_{m=1}^{n}\Delta_m W \widetilde{F}_{n,m}^{(\ell)} + G_n^{(\ell)}, \tag{3.102}$$

where $F_{n,m}^{(\ell)}$, $\widetilde{F}_{n,m}^{(\ell)}$, $G_n^{(\ell)}$ are deterministic and of the forms

$$\begin{aligned}
F_{n,m}^{(\ell)} &= -\tau^2 \sum_{j=n+1}^{N-1}\sum_{l=m}^{j-1} A_0^{2j-l-n} f_{l,m}^{(\ell)} - \alpha\tau \sum_{l=m}^{N-1} A_0^{2N-l-n} f_{l,m}^{(\ell)} \\
&\qquad\qquad\qquad\qquad m=1,2,\cdots,n-1, \\
F_{n,n}^{(\ell)} &= -\tau^2 \sum_{j=n+1}^{N-1}\Big[\sum_{l=n}^{j-1} A_0^{2j-l-n} f_{l,n}^{(\ell)} + \sum_{l=n+1}^{j-1} A_0^{2j-l-n} \sum_{m=n+1}^{l} f_{l,m}^{(\ell)}\Big] \\
&\quad -\alpha\tau\Big[\sum_{l=n}^{N-1} A_0^{2N-l-n} f_{l,n}^{(\ell)} + \sum_{l=n+1}^{N-1} A_0^{2N-l-n} \sum_{m=n+1}^{l} f_{l,m}^{(\ell)}\Big], \\
\widetilde{F}_{n,m}^{(\ell)} &= -\tau^2 \sum_{j=n+1}^{N-1}\sum_{l=m}^{j-1} A_0^{2j-l-n} \widetilde{f}_{l,m}^{(\ell)} - \alpha\tau \sum_{l=m}^{N-1} A_0^{2N-l-n} \widetilde{f}_{l,m}^{(\ell)} \\
&\quad -\tau \sum_{j=n+1}^{N-1} A_0^{2j-m-n+1}\Pi_h\sigma(t_{m-1}) - \alpha A_0^{2N-m-n+1}\Pi_h\sigma(t_{m-1}) \\
&\qquad\qquad\qquad\qquad m=1,2,\cdots,n, \\
G_n^{(\ell)} &= -\tau \sum_{j=n+1}^{N-1} A_0^{2j-n}\Pi_h x - \alpha A_0^{2N-n}\Pi_h x - \tau^2 \sum_{j=n+1}^{N-1}\sum_{l=0}^{j-1} A_0^{2j-l-n} g_l^{(\ell)} \\
&\quad -\alpha\tau \sum_{l=0}^{N-1} A_0^{2N-l-n} g_l^{(\ell)}.
\end{aligned} \tag{3.103}$$

Moreover, $f_{n,m}^{(\ell+1)}$, $\widetilde{f}_{n,m}^{(\ell+1)}$, $g_n^{(\ell+1)}$ in $U_{h\tau}^{(\ell+1)}(t_n)$ for $\ell \in \mathbb{N}_0$ can be derived recursively by

$$\begin{aligned}
f_{n,m}^{(\ell+1)} &= \Big(1-\frac{1}{\kappa}\Big) f_{n,m}^{(\ell)} + \frac{1}{\kappa} F_{n,m}^{(\ell)}, \\
\widetilde{f}_{n,m}^{(\ell+1)} &= \Big(1-\frac{1}{\kappa}\Big) \widetilde{f}_{n,m}^{(\ell)} + \frac{1}{\kappa} \widetilde{F}_{n,m}^{(\ell)} \qquad m=1,2,\cdots,n, \\
g_n^{(\ell+1)} &= \Big(1-\frac{1}{\kappa}\Big) g_n^{(\ell)} + \frac{1}{\kappa} G_n^{(\ell)}.
\end{aligned}$$

Proof For simplicity, in the proof we rewrite $X_{h\tau}^{(\ell)}(t_n)$, $Y_{h\tau}^{(\ell)}(t_n)$, $U_{h\tau}^{(\ell)}(t_n)$, $\Pi_h\sigma(t_n)$ as $X_n^{(\ell)}$, $Y_n^{(\ell)}$, $U_n^{(\ell)}$, σ_n. We argue by induction to derive the assertion. Suppose that

$$U_n^{(\ell)} = \sum_{m=1}^{n}\prod_{i=1}^{m}(1+\Delta_i W) f_{n,m}^{(\ell)} + \sum_{m=1}^{n}\Delta_m W \widetilde{f}_{n,m}^{(\ell)} + g_n^{(\ell)} \quad n = 0, 1, \cdots, N-1\,, \tag{3.104}$$

with deterministic $f_{n,m}^{(\ell)}$, $\widetilde{f}_{n,m}^{(\ell)}$, $g_n^{(\ell)}$.

Based on (3.100), we can deduce that

$$X_{n+1}^{(\ell)} = A_0^{n+1}\Pi_h x + \tau\sum_{j=0}^{n} A_0^{n+1-j} U_j^{(\ell)} + \sum_{j=0}^{n} A_0^{n+1-j}\sigma_j\Delta_{j+1}W\,. \tag{3.105}$$

Then inserting (3.105) into $Y_n^{(\ell)}$ of step 2 in Algorithm 3.9, we have

$$\begin{aligned} Y_n^{(\ell)} &= -\tau\mathbb{E}^{t_n}\Big[\sum_{j=n+1}^{N-1} A_0^{j-n}X_j^{(\ell)}\Big] - \alpha\mathbb{E}^{t_n}\big[A_0^{N-n}X_N^{(\ell)}\big] \\ &= -\tau\sum_{j=n+1}^{N-1} A_0^{2j-n}\Pi_h x - \alpha A_0^{2N-n}\Pi_h x \\ &\quad -\tau\sum_{j=n+1}^{N-1}\tau\sum_{l=0}^{j-1} A_0^{2j-l-n}\mathbb{E}^{t_n}\big[U_l^{(\ell)}\big] - \alpha\tau\sum_{l=0}^{N-1} A_0^{2N-l-n}\mathbb{E}^{t_n}\big[U_l^{(\ell)}\big] \\ &\quad -\sum_{j=n+1}^{N-1}\tau\sum_{l=0}^{j-1} A_0^{2j-l-n}\mathbb{E}^{t_n}\big[\sigma_l\Delta_{l+1}W\big] - \alpha\sum_{l=0}^{N-1} A_0^{2N-l-n}\mathbb{E}^{t_n}\big[\sigma_l\Delta_{l+1}W\big] \\ &=: -\tau\sum_{j=n+1}^{N-1} A_0^{2j-n}\Pi_h x - \alpha A_0^{2N-n}\Pi_h x - \sum_{i=1}^{4} I_i\,. \end{aligned} \tag{3.106}$$

Now we compute I_i for $1 \le i \le 4$. By the assumption for $U_l^{(\ell)}$, if $l \le n$ we have

$$\mathbb{E}^{t_n}\big[U_l^{(\ell)}\big] = U_l^{(\ell)}\,, \tag{3.107}$$

and if $l > n$, then by (3.104)

$$\begin{aligned} \mathbb{E}^{t_n}\big[U_l^{(\ell)}\big] &= \sum_{m=1}^{n-1}\prod_{i=1}^{m}(1+\Delta_i W) f_{l,m}^{(\ell)} + \prod_{i=1}^{n}(1+\Delta_i W)\Big[f_{l,n}^{(\ell)} + \sum_{m=n+1}^{l} f_{l,m}^{(\ell)}\Big] \\ &\quad + \sum_{m=1}^{n}\Delta_m W\widetilde{f}_{l,m}^{(\ell)} + g_l^{(\ell)}\,. \end{aligned} \tag{3.108}$$

Since $\sigma(\cdot)$ is deterministic, we have

$$\mathbb{E}^{t_n}\big[\sigma_l \Delta_{l+1} W\big] = \begin{cases} \sigma_l \Delta_{l+1} W & l \le n-1\,, \\ 0 & l \ge n\,. \end{cases} \tag{3.109}$$

For I_1, by (3.107)–(3.108), we then arrive at

$$\begin{aligned} I_1 &= \tau^2 \sum_{j=n+1}^{N-1} \sum_{l=0}^{n} A_0^{2j-l-n} \Big[\sum_{m=1}^{l} \prod_{i=1}^{m} (1+\Delta_i W) f_{l,m}^{(\ell)} + \sum_{m=1}^{l} \Delta_m W \widetilde{f}_{l,m}^{(\ell)} + g_l^{(\ell)} \Big] \\ &+ \tau^2 \sum_{j=n+1}^{N-1} \sum_{l=n+1}^{j-1} A_0^{2j-l-n} \Big\{ \sum_{m=1}^{n-1} \prod_{i=1}^{m} (1+\Delta_i W) f_{l,m}^{(\ell)} \\ &\quad + \prod_{i=1}^{n} (1+\Delta_i W) \Big[f_{l,n}^{(\ell)} + \sum_{m=n+1}^{l} f_{l,m}^{(\ell)} \Big] + \sum_{m=1}^{n} \Delta_m W \widetilde{f}_{l,m}^{(\ell)} + g_l^{(\ell)} \Big\}\,. \end{aligned}$$

By changing the order of the summation, we continue to get

$$\begin{aligned} &= \sum_{m=1}^{n-1} \prod_{i=1}^{m} (1+\Delta_i W) \tau^2 \sum_{j=n+1}^{N-1} \sum_{l=m}^{j-1} A_0^{2j-l-n} f_{l,m}^{(\ell)} \\ &+ \prod_{i=1}^{n} (1+\Delta_i W) \tau^2 \sum_{j=n+1}^{N-1} \Big[\sum_{l=n}^{j-1} A_0^{2j-l-n} f_{l,n}^{(\ell)} + \sum_{l=n+1}^{j-1} A_0^{2j-l-n} \sum_{m=n+1}^{l} f_{l,m}^{(\ell)} \Big] \\ &+ \sum_{m=1}^{n} \Delta_m W \tau^2 \sum_{j=n+1}^{N-1} \sum_{l=m}^{j-1} A_0^{2j-l-n} \widetilde{f}_{l,m}^{(\ell)} + \tau^2 \sum_{j=n+1}^{N-1} \sum_{l=0}^{j-1} A_0^{2j-l-n} g_l^{(\ell)}\,. \end{aligned}$$

Similarly, for I_2, we have

$$\begin{aligned} I_2 &= \sum_{m=1}^{n-1} \prod_{i=1}^{m} (1+\Delta_i W) \alpha\tau \sum_{l=m}^{N-1} A_0^{2N-l-n} f_{l,m}^{(\ell)} \\ &+ \prod_{i=1}^{n} (1+\Delta_i W) \alpha\tau \Big[\sum_{l=n}^{N-1} A_0^{2N-l-n} f_{l,n}^{(\ell)} + \sum_{l=n+1}^{N-1} A_0^{2N-l-n} \sum_{m=n+1}^{l} f_{l,m}^{(\ell)} \Big] \\ &+ \sum_{m=1}^{n} \Delta_m W \alpha\tau \sum_{l=m}^{N-1} A_0^{2N-l-n} \widetilde{f}_{l,m}^{(\ell)} + \alpha\tau \sum_{l=0}^{N-1} A_0^{2N-l-n} g_l^{(\ell)}\,. \end{aligned}$$

For I_3 and I_4, based on (3.109), we can conclude that

$$I_3 = \sum_{j=n+1}^{N-1} \tau \sum_{l=0}^{n-1} A_0^{2j-l-n} \sigma_l \Delta_{l+1} W = \sum_{m=1}^{n} \Delta_m W \tau \sum_{j=n+1}^{N-1} A_0^{2j-m-n+1} \sigma_{m-1}\,,$$

and

$$I_4 = \sum_{l=0}^{n-1} \alpha A_0^{2N-l-n} \sigma_l \Delta_{l+1} W = \sum_{m=1}^{n} \Delta_m W \alpha A_0^{2N-m-n+1} \sigma_{m-1} \,.$$

Finally, a combination of (3.106), the above computation of I_1 through I_4 and (3.103) then yields the assertion (3.102). That completes the proof. □

Remark 3.6 **(i)** From Theorem 3.11, we know that if $f_{\cdot,\cdot}^{(0)} = 0$, then $F_{\cdot,\cdot}^{(\ell)} = 0$ for any $\ell \in \mathbb{N}_0$.

(ii) When $\beta \neq 0$, (3.105) turns to

$$\begin{aligned} X_{h\tau}^{(\ell)}(t_{n+1}) &= A_0^{n+1} \prod_{j=1}^{n+1} (1+\beta\Delta_j W)\Pi_h x + \tau \sum_{j=0}^{n} A_0^{n+1-j} \prod_{k=j+2}^{n+1} (1+\beta\Delta_k W) U_{h\tau}^{(\ell)}(t_j) \\ &+ \sum_{j=0}^{n} A_0^{n+1-j} \prod_{k=j+2}^{n+1} (1+\beta\Delta_k W)\Pi_h \sigma(t_j)\Delta_{j+1} W \,. \end{aligned}$$

Then to compute $Y_{h\tau}^{(\ell)}$ (see step 2 of Algorithm 3.9), we have to deal with terms like

$$\mathbb{E}^{t_n}\Big[\prod_{k=n+2}^{j} (1+\beta\Delta_k W) U_l^{(\ell)} \Big],$$

for which it is impossible to derive the form as

$$\sum_{m=1}^{n} \prod_{i=1}^{m} (1+\Delta_i W) \bar{f}_{n,m}^{(\ell)} + \sum_{m=1}^{n} \Delta_m W \widehat{f}_{n,m}^{(\ell)} + \bar{g}_n^{(\ell)} \,,$$

with deterministic $\bar{f}_{n,m}^{(\ell)}$, $\widehat{f}_{n,m}^{(\ell)}$, $\bar{g}_n^{(\ell)}$.

In the sequel, we report on computational studies for the following problem, where the state equation is

$$\begin{cases} \mathrm{d}X(t) = \big[\Delta X(t) + U(t)\big]\mathrm{d}t + \sum_{i=1}^{m} \big[\beta_i X(t) + \sigma_i(t)\big]\mathrm{d}W_i(t) \quad t \in (0,T]\,, \\ X(0) = x \in \mathbb{H}_0^1 \cap \mathbb{H}^2 \,, \end{cases} \tag{3.110}$$

and a given function $\widetilde{X}(\cdot) : [0,T] \times D \to \mathbb{R}$ is in the cost functional

$$\mathcal{J}\big(U(\cdot)\big) = \frac{1}{2}\mathbb{E}\Big[\int_0^T \|X(t) - \widetilde{X}(t)\|^2 + \|U(t)\|^2 \,\mathrm{d}t + \alpha \|X(T) - \widetilde{X}(T)\|^2\Big], \tag{3.111}$$

where W is a $\mathbb{R}^m$-valued Brownian motion. For the following example, we use Algorithm 3.9 (in modified form), where step 2 is (3.102). These computations are done on a Dual Xeon 6242 Workstation with 768 GB RAM.

Example 3.1 Consider additive noise in (3.110), i.e., all $\beta_i = 0$. We set

$$D = (0,1)\,, \quad T = 0.4\,, \quad \alpha = 1\,, \quad m = 2\,, \quad \beta_i(x) = 0\,,$$
$$\sigma_i(t,x) = \frac{t}{i}\sin(\pi x)\,, \quad \widetilde{X}(t,x) = (2-8t)\sin(\pi x)\,, \quad X(0,x) = 2\sin(\pi x)\,.$$

Figure 3.2 (left) displays a decay of $\ell \mapsto \mathcal{J}_{h,\tau}\big(U_{h\tau}^{(\ell)}\big)$ for growing iterations ℓ, where $(\tau, h) = (0.1 \times \frac{1}{4}, \frac{1}{25})$, and $\mathtt{M} = 1000$ Monte Carlo simulations. The plot on the right-hand side shows experimentally strong convergence rates, for which the CPU time took about 10.5 days (parameters $\tau = 0.1 \times \frac{1}{2^4}$, $h = \frac{1}{25}$, $\ell = 10$, $\mathtt{M} = 1000$ were chosen to compute a reference solution).

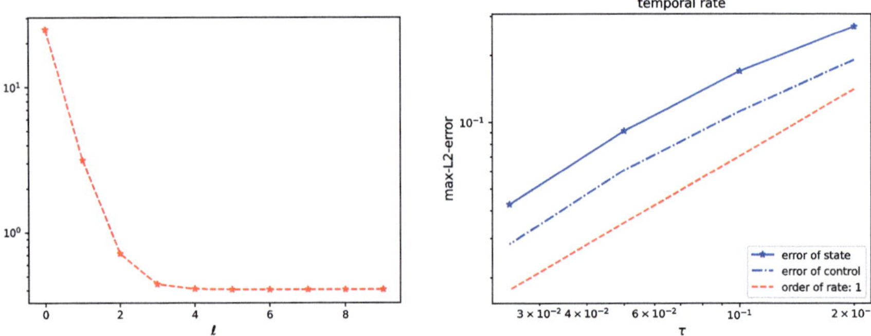

Fig. 3.2 Example 3.1: plot of $\ell \mapsto \mathcal{J}_{h,\tau}\big(U_{h\tau}^{(\ell)}\big)$ (left). Temporal rates of convergence (right); for discretization parameters $h = \frac{1}{25}$, $\tau = 0.2, 0.1, 0.1 \times \frac{1}{2}, 0.1 \times \frac{1}{2^2}$, for $\ell = 10$ gradient steps, and $\mathtt{M} = 1000$

3.4.2 A Regression Estimator for Conditional Expectations

The strategy in this section is to simulate Problem $\mathbf{(SLQ)}_{h\tau}$ for general $\beta \in \mathbb{R}$ via Algorithm 3.9. For fixed $\ell \in \mathbb{N}_0$ in Algorithm 3.9, step 2 has the form $Y_{h\tau}^{(\ell)}(t_n) = \mathbb{E}^{t_n}\big[\Theta_{h\tau}^{(\ell)}(t_n)\big]$, which involves a conditional expectation. We approximate it by a *'data dependent'* regression estimator (see e.g. [21, Chap. 13]) which is able to cope with the high dimensionality $1 \ll \mathfrak{d} \triangleq \dim(\mathbb{V}_h)$ of the involved state space.

As will be outlined below, related data dependent partitionings $\mathscr{P}_{\mathtt{M}}^{n,(\ell)}$ of $\mathbb{R}^{\mathfrak{d}}$ will automatically be generated, with finer cells at places where the regression function (see below) changes fastly, and coarser ones elsewhere. This nonlinear approximation approach significantly reduces complexities of simulations, in particular when $\mathfrak{d}$ is large—and where well-known *uniform* partitioning strategies for low-dimensional BSDEs loose their competitivity.

The regression estimator from [5] in this section may be considered as a modification of the least squares Monte Carlo method (see e.g. [21, 22]) to make accessible the simulation of high-dimensional BSDEs as required in step 2 of Algorithm 3.9. Choose $\mathtt{M}, \mathtt{R} \in \mathbb{N}$, with $\mathtt{R} \ll \mathtt{M}$. We use the following tools to simulate $\mathbb{E}^{t_n}\big[\Theta_{h\tau}^{(\ell)}(t_n)\big]$:

(a) **(algebraic reformulation)** Make use of the finite element basis functions $\{\phi_{h,j}\}_{j=1}^{\mathfrak{d}}$ of $\mathbb{V}_h$ to represent appearing $\mathbb{V}_h$-valued $Y_{h\tau}^{\cdot,(\ell)}$ resp. $X_{h\tau}^{\cdot,(\ell)}$ by $\mathbb{R}^{\mathfrak{d}}$-valued $\mathfrak{Y}^{\cdot,(\ell)}$ resp. $\mathfrak{X}^{\cdot,(\ell)}$, with related entries $\mathfrak{y}_j^{\cdot,(\ell)}$ resp. $\mathfrak{x}_j^{\cdot,(\ell)}$ via

$$\mathfrak{X}^{\cdot,(\ell)} = \sum_{j=1}^{\mathfrak{d}} \mathfrak{x}_j^{\cdot,(\ell)} \phi_{h,j} \qquad \text{and} \qquad \mathfrak{Y}^{\cdot,(\ell)} = \sum_{j=1}^{\mathfrak{d}} \mathfrak{y}_j^{\cdot,(\ell)} \phi_{h,j} \,.$$

(b) **($\mathtt{M}$-sample $\mathrm{D}_{\mathtt{M},j}^{n,(\ell)}$)** Let $1 \le j \le \mathfrak{d}$; the regression function $\mathcal{Y}_j^{n,(\ell)} : \mathbb{R}^{\mathfrak{d}} \to \mathbb{R}$ satisfies

$$\mathfrak{y}_j^{n,(\ell)} = \mathcal{Y}_j^{n,(\ell)}\big(\mathfrak{X}^{n,(\ell)}\big) \,,$$

which we approximate by the estimator $\widehat{\mathcal{Y}}_j^{n,(\ell)}$ in (d) below. Therefore, let

$$\mathrm{D}_{\mathtt{M},j}^{n,(\ell)} \triangleq \big\{\big(\mathfrak{X}_{\mathtt{m}}^{n,(\ell)}, \theta_{\mathtt{m},j}^{n,(\ell)}\big)\big\}_{\mathtt{m}=1}^{\mathtt{M}} \qquad (1 \le n \le N)$$

be an $\mathtt{M}$-sample, with $\big(\mathfrak{X}_{\mathtt{m}}^{n,(\ell)}, \theta_{\mathtt{m},j}^{n,(\ell)}\big) \sim \big(\mathfrak{X}^{n,(\ell)}, \theta_j^{n,(\ell)}\big)$, and $\theta_j^{n,(\ell)}$ the j-th entry of the $\mathbb{R}^{\mathfrak{d}}$-valued representation of $\Theta_{h\tau}^{(\ell)}(t_n)$.

(c) **(data dependent partition $\mathscr{P}_{\mathtt{M}}^{n,(\ell)}$)** Use (the first entries of) $\mathrm{D}_{\mathtt{M},j}^{n,(\ell)}$ to generate a partition (see also [21, Chap. 13])

$$\mathscr{P}_{\mathtt{M}}^{n,(\ell)} = \big\{\mathscr{A}_{\mathtt{r}}^{n,(\ell)}\big\}_{\mathtt{r}=1}^{\mathtt{R}}$$

of $\mathbb{R}^{\mathfrak{d}}$ into statistically equivalent cells $\mathscr{A}_{\mathtt{r}}^{n,(\ell)}$ via the BTC strategy in [5, p. A2741].

(d) **(estimator $\widehat{\mathcal{Y}}_j^{n,(\ell)}$)** Use $\mathrm{D}_{\mathtt{M},j}^{n,(\ell)}$ to compute $\widehat{\mathcal{Y}}_j^{n,(\ell)} : \mathbb{R}^{\mathfrak{d}} \to \mathbb{R}$ via

$$\widehat{\mathcal{Y}}_j^{n,(\ell)}(\cdot) = \sum_{\mathtt{r}=1}^{\mathtt{R}} \frac{\sum_{\mathtt{m}=1}^{\mathtt{M}} \theta_{\mathtt{m},j}^{n,(\ell)} \cdot \chi_{\{\mathfrak{X}_{\mathtt{m}}^{n,(\ell)} \in \mathscr{A}_{\mathtt{r}}^{n,(\ell)}\}}}{\sum_{\mathtt{m}=1}^{\mathtt{M}} \chi_{\{\mathfrak{X}_{\mathtt{m}}^{n,(\ell)} \in \mathscr{A}_{\mathtt{r}}^{n,(\ell)}\}}} \chi_{\mathscr{A}_{\mathtt{r}}}(\cdot) \,.$$

We refer to [5] for more algorithmic details.

The following example compares the efficiencies of the regression estimator in step 2 of Algorithm 3.9 with the exact computation (3.102) in the case of Problem **(SLQ)** (3.111) subject to (3.110)—provided that the noise is additive; and it discusses computations via the regression estimator in the case of multiplicative noise.[2]

---

[2] We are grateful to F. Merle (U Tübingen) who provided the simulations in Examples 3.1 and 3.2.

Example 3.2 1. Consider additive noise case in (3.110), i.e., all $\beta_i = 0$. We set

$$D = (0, 1)\,, \quad T = 0.2\,, \quad \alpha = 1\,, \quad m = 5\,, \quad \beta_i(x) = 0\,,$$
$$\sigma_i(t, x) = 2\sin(i\pi x)\,, \quad \widetilde{X}(t, x) = \frac{2(T - t)}{T}\sin(\pi x)\,, \quad X(0, x) = 2\sin(\pi x)\,.$$

We choose $h = \frac{1}{25}$, $\tau = 0.1 \times \frac{1}{2^3}$, $\kappa = 1 + \alpha T + T^2$, and $\ell = 5$ gradient descent iterations. Figure 3.3 displays one trajectory of optimal state (left) and optimal control (middle). Its simulation is via Algorithm 3.9 with the exact computation (3.102), where the CPU-time was 3 s.

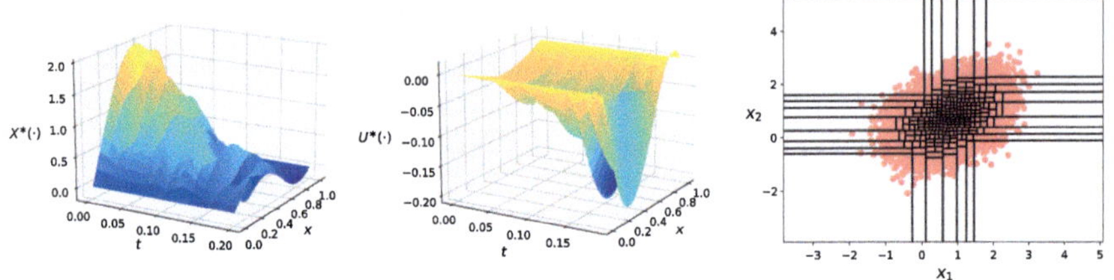

Fig. 3.3 Example 3.2: Single trajectory for $\big(X_{h\tau}^{(\ell)}(\cdot), U_{h\tau}^{(\ell)}(\cdot)\big)$ in part 1 of the example (left and middle). Plot of $\mathscr{P}_{\mathrm{M}}^{n,(\ell)}$ if Algorithm 3.9 together with the regression estimator from [5] is used (right)

2. Now we use Algorithm 3.9 together with the regression estimator from [5] sketched above when we consider linear noise in (3.110):

$$D = (0, 1)\,, \quad T = 0.2\,, \quad \alpha = 1\,, \quad m = 5\,, \quad \beta_i(x) = 2\,,$$
$$\sigma_i(t, x) = 0\,, \quad \widetilde{X}(t, x) = \frac{2(T - t)}{T}\sin(\pi x)\,, \quad X(0, x) = 2\sin(\pi x)\,.$$

The numerical parameters are as in 1. Figure 3.4 displays one trajectory of optimal state (left) and optimal control (middle). Its simulation is via Algorithm 3.9 with the regression estimator, and the CPU-time was much larger—it was 31552 s!

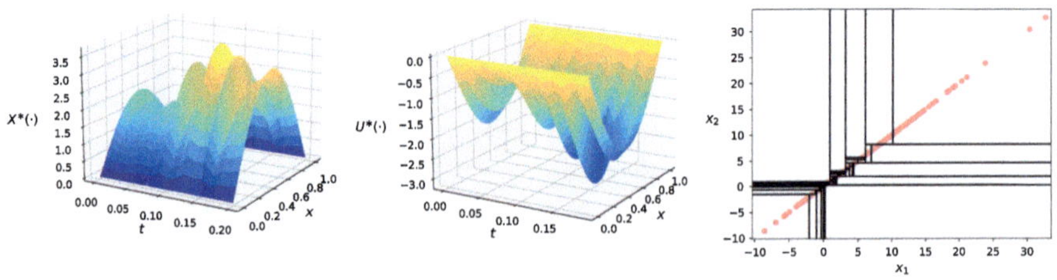

Fig. 3.4 Example 3.2: Single trajectory for $\big(X_{h\tau}^{(\ell)}(\cdot), U_{h\tau}^{(\ell)}(\cdot)\big)$ in part 2 (left and middle). Plot of $\mathscr{P}_{\mathrm{M}}^{n,(\ell)}$ if Algorithm 3.9 together with the regression estimator from [5] is used (right)

3. The plots on the right side in Figs. 3.3 and 3.4 give the data dependent partitions $\mathscr{P}_{\mathtt{M}}^{n,(\ell)}$ (at time $t_n = \frac{T}{2}$) in the case of additive resp. linear noise: here for $h = \frac{1}{3}$ to allow visualization in 2D (with $\mathtt{M} = 2^{14}$, and number of cells $\mathtt{R} = 2^8$).

References

1. R. Kruse, Strong and weak approximation of semilinear stochastic evolution equations. *Lecture Notes in Mathematics*, vol. 2093 (Springer, Cham, 2014)
2. X. Wang, Strong convergence rates of the linear implicit Euler method for the finite element discretization of SPDEs with additive noise. IMA J. Numer. Anal. **37**, 965–984 (2017)
3. A. Tambue, J.D. Mukam, Strong convergence of the linear implicit Euler method for the finite element discretization of semilinear SPDEs driven by multiplicative or additive noise. Appl. Math. Comput. **346**, 23–40 (2019)
4. N. El Karoui, S. Peng, M.C. Quenez, Backward stochastic differential equations in finance. Math. Finance **7**, 1–71 (1997)
5. T. Dunst, A. Prohl, The forward-backward stochastic heat equation: numerical analysis and simulation. SIAM J. Sci. Comput. **38**, A2725–A2755 (2016)
6. É. Pardoux, S. Peng, Adapted solution of a backward stochastic differential equation. Syst. Control Lett. **14**, 55–61 (1990)
7. J. Zhang, A numerical scheme for BSDEs. Ann. Appl. Probab. **14**, 459–488 (2004)
8. Y. Hu, D. Nualart, X. Song, Malliavin calculus for backward stochastic differential equations and application to numerical solutions. Ann. Appl. Probab. **21**, 2379–2423 (2011)
9. Y. Wang, A semidiscrete Galerkin scheme for backward stochastic parabolic differential equations. Math. Control Relat. Fields **6**, 489–515 (2016)
10. M. Hinze, R. Pinnau, M. Ulbrich, S. Ulbrich, Optimization with PDE constraints. *Mathematical Modelling: Theory and Applications*, vol. 23 (Springer, New York, 2009)
11. A. Prohl, Y. Wang, Strong rates of convergence for a space-time discretization of the backward stochastic heat equation, and of a linear-quadratic control problem for the stochastic heat equation. ESAIM Control Optim. Calc. Var. **27**, Paper No. 54, 30 (2021)
12. B. Li, Q. Zhou, Discretization of a distributed optimal control problem with a stochastic parabolic equation driven by multiplicative noise. J. Sci. Comput. **87**, Paper No. 68, 37 (2021)
13. Q. Lü, P. Wang, Y. Wang, X. Zhang, Chapter 6—numerics for stochastic distributed parameter control systems: a finite transposition method, in *Numerical Control: Part A*, ed. by E. Trélat, E. Zuazua, vol. 23 of Handbook of Numerical Analysis (Elsevier, 2022), pp. 201–232
14. A. Prohl, Y. Wang, Strong error estimates for a space-time discretization of the linear-quadratic control problem with the stochastic heat equation with linear noise. IMA J. Numer. Anal. **42**, 3386–3429 (2022)
15. Y. Wang, Strong error estimates for the space-time discretization of a stochastic linear quadratic control problem with control in the diffusion. SIAM J. Control Optim. **63**, 3328–3355 (2025)
16. J. Yong, X.Y. Zhou, Stochastic controls: hamiltonian systems and HJB equations. *Applications of Mathematics (New York)*, vol. 43. (Springer, New York, 1999)
17. D. Nualart, The Malliavin calculus and related topics. *Probability and its Applications (New York)*, 2nd ed. (Springer, Berlin, 2006)
18. Y. Nesterov, Introductory lectures on convex optimization. *Applied Optimization*, vol. 87 (Kluwer Academic Publishers, Boston, MA, 2004)
19. S.I. Kabanikhin, Inverse and ill-posed problems. *Inverse and Ill-posed Problems Series*, vol. 55 (Walter de Gruyter GmbH & Co. KG, Berlin, 2012)
20. A. Chaudhary, F. Merle, A. Prohl, Y. Wang, An efficient discretization to simulate the solution of linear-quadratic stochastic boundary control problem. IMA J. Numer. Anal. to appear

21. L. Györfi, M. Kohler, A. Krzyżak, H. Walk, A distribution-free theory of nonparametric regression. *Springer Series in Statistics* (Springer, New York, 2002)
22. E. Gobet, P. Turkedjiev, Linear regression MDP scheme for discrete backward stochastic differential equations under general conditions. Math. Comput. **85**, 1359–1391 (2016)

Chapter 4
Discretization Based on the Closed-Loop Approach

As has been reviewed in Sect. 2.3, the optimal control $U^*(\cdot)$ of Problem **(SLQ)** admits the feedback law (1.11) with the help of $\mathcal{P}(\cdot)$ and $\eta(\cdot)$ (see (1.9) and (1.10)). In comparison to Chap. 3, both terminal problems on (1.9) and (1.10) are deterministic—a relevant property that we may now benefit from in a numerical discretization of Problem **(SLQ)**: a canonical strategy therefore is now to approximate $U^*(\cdot)$ by first discretizing (1.9) and (1.10) with the help of tools from deterministic numerics; and to finally insert it into SPDE (1.2) thanks to (1.11), and to then solve it numerically to get an approximation of $X^*(\cdot)$.

As a consequence, to properly discretize the stochastic Riccati equation (1.9) is the key step within this program, and to derive related optimal error estimates is the main aim in this chapter. In fact, it is an open problem even in the context of deterministic control (i.e., $\beta = 0$ and $\sigma(\cdot) = 0$ in Problem **(SLQ)**) whether the implicit Euler for the *deterministic Riccati equation* performs with optimal rates: the reason for it is that an applied discretization can spoil the specific role that the continuous Riccati equation plays within the (limiting) control problem. To preserve this role for the constructed discretization is of primary relevance in this chapter. By adopting an idea from [1], we develop a proper discretization scheme of (1.9)—which, in effect, differs from a standard time numerical integrator for (1.9); see (4.20) and (4.68). In the second step, we apply a perturbation argument, where we leave the ground of optimization to arrive at a modified scheme whose complexity is comparable to standard numerical integrators for (1.9). The benefit of this construction strategy for the related scheme (4.99)–(4.100) is that we can show optimal rates of convergence for iterates to approximate $\big(X^*(\cdot), U^*(\cdot)\big)$.

Based on LQ theory, the error analysis for different spatio-temporal discretization schemes of the Riccati equation is given in Sect. 4.2; see Theorems 4.1 (for spatial discretization), 4.2, 4.3 (for temporal discretization in $\|\cdot\|_{\mathcal{L}(\mathbb{L}^2)}$) under some assumption on β (see assumption **(H)** in Sect. 4.1.2), and Theorems 4.5, 4.6 (for temporal discretization in $\|\cdot\|_{\mathcal{L}(\mathbb{H}_0^1;\mathbb{L}^2)}$) *without* any smallness assumption on β.

© The Author(s), under exclusive license to Springer Nature Singapore Pte Ltd. 2026
A. Prohl and Y. Wang, *Numerical Methods for Optimal Control Problems with SPDEs*, SpringerBriefs on PDEs and Data Science,
https://doi.org/10.1007/978-981-95-4469-1_4

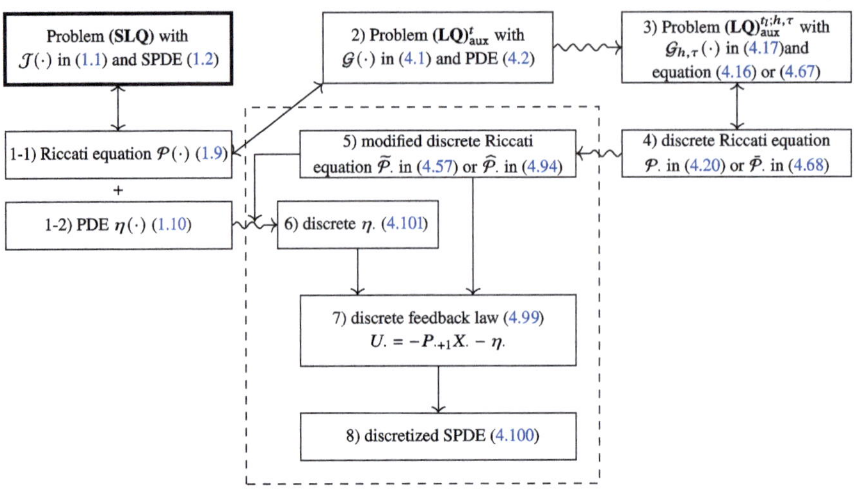

Fig. 4.1 Construction of the discretization scheme in dashed box to approximate Problem **(SLQ)**; used wiggly arrows indicate order preserving approximation or perturbation

In Sect. 4.3, a mixed strategy to numerically solve Problem **(SLQ)** is proposed, which uses the 'open-loop approach' for spatial discretization, and then the 'closed-loop approach' for temporal discretization. This mixed approach improves the results in [2–4] which adopts the 'open-loop approach', and in [5] where methods were only taken from the 'closed-loop approach' at the stage of the numerical analysis.

This chapter starts with Sect. 4.1, where we introduce a family of LQ problems (Problem $(\mathbf{LQ})^t_{\mathtt{aux}}$) related to Problem **(SLQ)**, which will play a prominent role in the numerical analysis. In fact, the *stochastic* Riccati equation for Problem **(SLQ)** coincides with the *deterministic* Riccati equation for Problem $(\mathbf{LQ})^t_{\mathtt{aux}}$, where the equation is a (deterministic) PDE instead, containing a shifted operator on a scale β if compared with the original diffusion operator in (1.2). Figure 4.1 shows the outline of the 'closed-loop approach'.

4.1 Discretization of the Riccati Equation

The *stochastic* Riccati equation (1.9) is related to Problem **(SLQ)**; but it can also be considered as a *deterministic* Riccati equation associated with Problem $(\mathbf{LQ})^{\cdot}_{\mathtt{aux}}$ below. We will detail this observation in this section, which will play a relevant role in the construction of a proper discretization scheme for (1.9)—if interpreted in the latter sense. In this chapter, we propose two different temporal discretization schemes for the Riccati equation (1.9) which are both consistent—i.e., its solutions take values in $\mathbb{S}_+(\mathbb{L}^2|_{\mathbb{V}_h})$—and two further schemes which are not; their construction again follows the strategy to 'first discretize, then optimize'. To state the main ideas, in this section we only detail the first consistent discretization; the others will be stated in part **b.** of Sect. 4.2.2, and in Sect. 4.2.3.

4.1.1 An Auxiliary LQ Problem for the Riccati Equation

We now define a family of *deterministic* LQ problems—where each of them is later referred to as Problem $(\mathbf{LQ})^t_{\text{aux}}$ for any $t \in [0, T)$—whose corresponding Riccati equation is (1.9) on $[t, T]$. To do that, we define a family of cost functionals first, which for any $t \in [0, T)$ and $z \in \mathbb{L}^2$ reads as

$$\mathcal{G}\big(t, z; u(\cdot)\big) = \frac{1}{2}\int_t^T \big[\|x(s)\|^2 + \|u(s)\|^2\big]\, \mathrm{d}s + \frac{\alpha}{2}\|x(T)\|^2 , \tag{4.1}$$

subject to the (controlled forward) *deterministic* PDE

$$\begin{cases} x'(s) = \Delta x(s) + \dfrac{\beta^2}{2} x(s) + u(s) & s \in (t, T] , \\ x(t) = z . \end{cases} \tag{4.2}$$

Note that the involved $\big(x(\cdot), u(\cdot)\big)$ in (4.1)–(4.2) are deterministic functions—as opposed to the related stochastic processes in (1.1)–(1.2).

Problem $(\mathbf{LQ})^t_{\text{aux}}$. Fix an initial pair $(t, z) \in [0, T) \times \mathbb{L}^2$. Find $u^*(\cdot) \in L^2(t, T; \mathbb{L}^2)$ such that

$$\mathcal{G}\big(t, z; u^*(\cdot)\big) = \inf_{u(\cdot) \in L^2(t,T;\mathbb{L}^2)} \mathcal{G}\big(t, z; u(\cdot)\big) =: V(t, z) . \tag{4.3}$$

Any $u^*(\cdot) \in L^2(t, T; \mathbb{L}^2)$ that satisfies (4.3) is called an *optimal control* for the given initial pair (t, z), with $x^*(\cdot)$ the corresponding *optimal state*; the tuple $\big(x^*(\cdot), u^*(\cdot)\big)$ will be referred to as an *optimal pair* for (t, z), and $V(\cdot, \cdot)$ is called the *value function* of Problem $(\mathbf{LQ})^t_{\text{aux}}$.

It is well-known that Problem $(\mathbf{LQ})^t_{\text{aux}}$ admits a unique optimal pair $\big(x^*(\cdot), u^*(\cdot)\big)$, and the optimal control $u^*(\cdot)$ admits the following state feedback representation:

$$u^*(s) = -\mathcal{P}(s)x^*(s) \qquad s \in [t, T] , \tag{4.4}$$

where $\mathcal{P}(\cdot)$ comes from the terminal problem (1.9). Furthermore, the value function $V(\cdot, \cdot)$ is explicitly presented by

$$V(t, z) = \frac{1}{2}\big(\mathcal{P}(t)z, z\big)_{\mathbb{L}^2} . \tag{4.5}$$

We refer the reader to [6, Chap. 17] for more details. Based on (4.3) and (4.5), for any admissible control $u(\cdot) \in L^2(t, T; \mathbb{L}^2)$ we may conclude that

$$\big(\mathcal{P}(t)z, z\big)_{\mathbb{L}^2} \leq 2\mathcal{G}\big(t, z; u(\cdot)\big),$$

which will be used frequently in Sect. 4.2.

4.1.2 Discretization of Problem $(\mathbf{LQ})^t_{\mathbf{aux}}$ and a Related Difference Riccati Equation

We derive the difference Riccati equation (4.20) with the help of two discretization schemes of Problem $(\mathbf{LQ})^t_{\mathrm{aux}}$. For abbreviation, we denote $\mathcal{A} = \Delta + \frac{\beta^2}{2}\mathbb{1}$ and shall make use of the following assumption:

Assumption (H): There exists a positive constant C_0 such that

$$\|\nabla\varphi\|^2 - \frac{\beta^2}{2}\|\varphi\|^2 \geq C_0\|\varphi\|^2 \qquad \forall\, \varphi \in \mathbb{H}^1_0. \tag{4.6}$$

This assumption constrains the strength of admissible noise, but is suitable for the first numerical analysis in this section and Sect. 4.2.2; it will, however, be weakened in Sect. 4.2.3. Under the coercivity assumption **(H)**, $\mathcal{A}$ generates an analytic semigroup $\widehat{E}(\cdot)$ with $\widehat{E}(t) = e^{t\mathcal{A}}$ for $t \geq 0$. Also, we can define $\mathcal{A}_h = \Delta_h + \frac{\beta^2}{2}\mathbb{1}_h : \mathbb{V}_h \to \mathbb{V}_h$, and consider its corresponding semigroup $\widehat{E}_h(\cdot)$, as well as $\mathcal{A}_0 = (\mathbb{1}_h - \tau\mathcal{A}_h)^{-1}$ and the related $\widehat{E}_{h,\tau}(\cdot)$. Similarly, we can define $\widehat{G}_h(\cdot)$, $\widehat{G}_\tau(\cdot)$, for notations see Sect. 2.1. Besides, Lemmata 2.1–2.4 still hold for $\mathcal{A}$.

a. Spatial discretization of Problem $(\mathbf{LQ})^t_{\mathrm{aux}}$. By virtue of $\mathcal{A}$, (1.9) can be rewritten as

$$\begin{cases} \mathcal{P}'(t) + \mathcal{A}\mathcal{P}(t) + \mathcal{P}(t)\mathcal{A} + \mathbb{1} - \mathcal{P}^2(t) = 0 & t \in [0, T], \\ \mathcal{P}(T) = \alpha\mathbb{1}, \end{cases} \tag{4.7}$$

and its spatial discretization by the finite element method is

$$\begin{cases} \mathcal{P}_h'(t) + \mathcal{A}_h\mathcal{P}_h(t) + \mathcal{P}_h(t)\mathcal{A}_h + \mathbb{1}_h - \mathcal{P}_h^2(t) = 0 & t \in [0, T], \\ \mathcal{P}_h(T) = \alpha\mathbb{1}_h. \end{cases} \tag{4.8}$$

By [7, Theorem 7.2], (4.8) admits a unique solution $\mathcal{P}_h(\cdot) \in C\big([0, T]; \mathbb{S}(\mathbb{L}^2|_{\mathbb{V}_h})\big)$. To measure the difference $\mathcal{P}(\cdot) - \mathcal{P}_h(\cdot)\Pi_h$, we introduce a family of auxiliary LQ problems.

Problem $(\mathrm{LQ})^{t;h}_{\mathrm{aux}}$. For any given $t \in [0, T)$ and $z \in \mathbb{L}^2$, search for $u_h^*(\cdot) \in L^2(t, T; \mathbb{L}^2)$ such that

$$\mathcal{G}_h\big(t, z; u_h^*(\cdot)\big) = \inf_{u_h(\cdot)\in L^2(t,T;\mathbb{L}^2)} \mathcal{G}_h\big(t, z; u_h(\cdot)\big) =: V_h(t, z), \tag{4.9}$$

where the cost functional is

$$\mathcal{G}_h\big(t, z; u_h(\cdot)\big) = \frac{1}{2}\int_t^T \big[\|x_h(s)\|^2 + \|u_h(s)\|^2\big]\,\mathrm{d}s + \frac{\alpha}{2}\|x_h(T)\|^2 \tag{4.10}$$

and the state variable $x_h(\cdot) \in C([t, T]; \mathbb{V}_h)$ satisfies

$$\begin{cases} x_h'(s) = \mathcal{A}_h x_h(s) + \Pi_h u_h(s) & s \in (t, T]\,, \\ x_h(t) = \Pi_h z\,. \end{cases} \tag{4.11}$$

Similar to Problem $\mathbf{(LQ)}^t_{\mathrm{aux}}$, Problem $\mathbf{(LQ)}^{t;h}_{\mathrm{aux}}$ has a unique optimal control $u_h^*(\cdot) \in L^2(0, T; \mathbb{L}^2)$, satisfying the following state feedback representation:

$$u_h^*(s) = -\mathcal{P}_h(s)x_h^*(s) \qquad s \in [t, T]\,; \tag{4.12}$$

and the value function $V_h(\cdot, \cdot)$ is explicitly presented by

$$V_h(t, z) = \frac{1}{2}\big(\mathcal{P}_h(t)\Pi_h z, z\big)_{\mathbb{L}^2}\,. \tag{4.13}$$

Based on (4.9) and (4.13), for any admissible control $u_h(\cdot) \in L^2(t, T; \mathbb{L}^2)$ we find that

$$\big(\mathcal{P}_h(t)\Pi_h z, z\big)_{\mathbb{L}^2} \le 2\mathcal{G}_h\big(t, z; u_h(\cdot)\big) \qquad \forall\, t \in [0, T]\,,$$

which will be used in Sect. 4.2.

Moreover, we can deduce that the optimal control $u_h^*(\cdot) \in L^2(t, T; \mathbb{V}_h)$. In fact, since $\Pi_h u_h^*(\cdot)$ is an admissible control, we conclude by (2.3) that

$$0 \le \mathcal{G}_h\big(t, z; \Pi_h u_h^*(\cdot)\big) - \mathcal{G}_h\big(t, z; u_h^*(\cdot)\big) = \frac{1}{2}\int_t^T \big[\|\Pi_h u_h^*(s)\|^2 - \|u_h^*(s)\|^2\big]\,\mathrm{d}s \le 0\,, \tag{4.14}$$

which yields $u_h^*(\cdot) = \Pi_h u_h^*(\cdot) \in L^2(t, T; \mathbb{V}_h)$.

b. Spatio-temporal discretization of Problem $\mathbf{(LQ)}^t_{\mathrm{aux}}$. For a given $t \in [0, T)$, we now discretize Problem $\mathbf{(LQ)}^{t;h}_{\mathrm{aux}}$ in time. Firstly, we need two spaces for state and control variables: for any $l = 0, 1, \ldots, N-1$,

$$\begin{aligned} \mathbb{X}(t_l, T) &\triangleq \Big\{x_\cdot \equiv \{x_n\}_{n=l}^{N-1} \;\Big|\; x_n \in \mathbb{V}_h \;\; \forall\, n = l, \cdots, N-1\,, \\ &\qquad\qquad \text{and } \tau \sum_{n=l}^{N-1} \|x_n\|^2 < \infty\Big\}\,, \\ \mathbb{U}(t_l, T) &\triangleq \Big\{u_\cdot \equiv \{u_n\}_{n=l}^{N-1} \;\Big|\; u_n \in \mathbb{L}^2 \;\; \forall\, n = l \cdots, N-1\,, \\ &\qquad\qquad \text{and } \tau \sum_{n=l}^{N-1} \|u_n\|^2 < \infty\Big\}\,. \end{aligned} \tag{4.15}$$

We endow the discrete state and control spaces with the following norms,

$$\|x.\|_{\mathbb{X}(t_l,T)} \triangleq \Big(\tau \sum_{n=l}^{N-1} \|x_n\|^2\Big)^{1/2}, \quad \|u.\|_{\mathbb{U}(t_l,T)} \triangleq \Big(\tau \sum_{n=l}^{N-1} \|u_n\|^2\Big)^{1/2}.$$

In this chapter, to deal with the difference equation, we use a subscript index for the time variable, such as x_n, u_n, and $x.$, $u.$.

By adopting the (semi-)implicit Euler method in time, we approximate system (4.11) and the cost functional (4.10) as follows: For any given $l = 0, 1, \ldots, N-1$, and $z \in \mathbb{L}^2$, the discrete state $x.$ satisfies

$$\begin{cases} x_{n+1} = \mathcal{A}_0 x_n + \tau \mathcal{A}_0 \Pi_h u_n & n = l, l+1, \ldots, N-1, \\ x_l = \Pi_h z, \end{cases} \tag{4.16}$$

where $\mathcal{A}_0 = \big(\mathbb{1}_h - \tau \mathcal{A}_h\big)^{-1}$, and

$$\mathcal{G}_{h,\tau}(t_l, z; u.) = \frac{1}{2}\big[\|x.\|^2_{\mathbb{X}(t_l,T)} + \|u.\|^2_{\mathbb{U}(t_l,T)}\big] + \frac{\alpha}{2}\|x_N\|^2. \tag{4.17}$$

A discretized version in time of the family of Problems $\mathbf{(LQ)}^{t;h}_{\text{aux}}$ is

Problem $\mathbf{(LQ)}^{t_l;h,\tau}_{\text{aux}}$. For any given $l = 0, 1, \ldots, N-1$ and $z \in \mathbb{L}^2$, search for $u^*_\cdot \in \mathbb{U}(t_l, T)$ such that

$$\mathcal{G}_{h,\tau}(t_l, z; u^*_\cdot) = \inf_{u.\in\mathbb{U}(t_l,T)} \mathcal{G}_{h,\tau}(t_l, z; u.) =: V_{h,\tau}(t_l, z).$$

Note that for Problems $\mathbf{(LQ)}^{t;h}_{\text{aux}}$ resp. $\mathbf{(LQ)}^{t_l;h,\tau}_{\text{aux}}$ the elements $x_h(\cdot) \in L^2(0,T;\mathbb{V}_h)$ resp. $x. \in \mathbb{X}(t_l, T)$ take values in $\mathbb{V}_h$, while $u_h(\cdot) \in L^2(0,T;\mathbb{L}^2)$ resp. $u. \in \mathbb{U}(t_l,T)$ take values in $\mathbb{L}^2$. This choice is used to prove convergence rates for the proposed discretization schemes of the Riccati equation; see Theorems 4.1 and 4.2. Actually, with the same argument as in (4.14), we can deduce that the optimal control $u^*_\cdot$ to Problem $\mathbf{(LQ)}^{t_l;h,\tau}_{\text{aux}}$ is $\mathbb{V}_h$-valued.

We call the solution of Problem $\mathbf{(LQ)}^{t_l;h,\tau}_{\text{aux}}$ its *optimal control* $u^*_\cdot \in \mathbb{U}(t_l, T)$, and the corresponding state $x^*_\cdot \in \mathbb{X}(t_l, T)$ an *optimal state*. The following result show the solvability of Problem $\mathbf{(LQ)}^{t_l;h,\tau}_{\text{aux}}$ and a *discrete feedback law* for the optimal control. We refer the reader to [8, Chap. 5] for deterministic systems, and to [9] for stochastic systems.

Lemma 4.1 *For any* $l = 0, 1, \ldots, N-1$ *and* $z \in \mathbb{L}^2$, *Problem* $\mathbf{(LQ)}^{t_l;h,\tau}_{\text{aux}}$ *admits a unique optimal control which enjoys the discrete state feedback law*

$$u^*_n = -\mathcal{K}_n^{-1}\mathcal{H}_n x^*_n \qquad n = l, l+1, \ldots, N-1. \tag{4.18}$$

Moreover, the value function satisfies

$$V_{h,\tau}(t_l, z) = \frac{1}{2}(\mathcal{P}_l \Pi_h z, z)_{\mathbb{L}^2} \qquad \forall z \in \mathbb{L}^2, \tag{4.19}$$

where $\mathcal{P}_\cdot \equiv \{\mathcal{P}_n\}_{n=0}^N \subset \mathbb{S}_+(\mathbb{L}^2|_{\mathbb{V}_h})$ *solves the following difference Riccati equation*

$$\begin{cases} \mathcal{P}_n = \mathcal{A}_0\mathcal{P}_{n+1}\mathcal{A}_0 + \tau\mathbb{1}_h - \tau\mathcal{H}_n\mathcal{K}_n^{-1}\mathcal{H}_n \quad n = 0, 1, \cdots, N-1\,, \\ \mathcal{P}_N = \alpha\mathbb{1}_h\,, \\ \mathcal{H}_n = \mathcal{A}_0\mathcal{P}_{n+1}\mathcal{A}_0\,, \quad \mathcal{K}_n = \mathbb{1}_h + \tau\mathcal{A}_0\mathcal{P}_{n+1}\mathcal{A}_0\,. \end{cases} \tag{4.20}$$

Proof We divide the proof into three steps. For simplicity, we set

$$K(x,u) = \frac{\tau}{2}\big[\|x\|^2 + \|u\|^2\big], \quad z_0 = \Pi_h z\,, \quad \mathcal{G}_N(x) = \frac{\alpha}{2}\|x\|^2\,,$$

and define recursively

$$\mathcal{G}_n(x) = \inf_{u\in\mathbb{L}^2}\big\{K(x,u) + \mathcal{G}_{n+1}(\mathcal{A}_0 x + \tau\mathcal{A}_0\Pi_h u)\big\} \qquad n = 0, 1, \ldots, N-1\,.$$

(1) Claim: For any $n = l, l+1, \ldots, N-1$, if a feedback law $u = F_n x$ minimizes $\mathcal{G}_n(x)$, then $(x_\cdot^*, u_\cdot^*)$ which is given by

$$\begin{cases} x_{n+1}^* = \mathcal{A}_0 x_n^* + \tau\mathcal{A}_0\Pi_h u_n^* \quad \text{with} \quad x_l^* = z_0\,, \\ u_n^* = F_n x_n^* \qquad n = l, l+1, \ldots, N-1 \end{cases}$$

is the optimal pair of Problem $(\mathbf{LQ})_{\text{aux}}^{t_l;h,\tau}$, and

$$\mathcal{G}_{h,\tau}(t_l, z; u_\cdot^*) = \mathcal{G}_l(z_0)\,. \tag{4.21}$$

In the following, we adopt a dynamic programming idea to derive this claim. By virtue of $K(\cdot,\cdot)$ and $\mathcal{G}_\cdot(\cdot)$, we can rewrite $\inf_{u_\cdot\in\mathbb{U}(t_l,T)}\mathcal{G}_{h,\tau}(t_l, z; u_\cdot)$ as

$$\begin{aligned} \inf_{u_\cdot\in\mathbb{U}(t_l,T)}\mathcal{G}_{h,\tau}(t_l, z; u_\cdot) &= \inf_{u_n\in\mathbb{L}^2,\, l\le n\le N-1}\Big\{\sum_{n=l}^{N-1} K(x_n, u_n) + \mathcal{G}_N(x_N)\Big\} \\ = \inf_{u_n\in\mathbb{L}^2,\, l\le n\le N-2}&\Big\{\sum_{n=l}^{N-2} K(x_n, u_n) + \inf_{u_{N-1}\in\mathbb{L}^2}\big\{K(x_{N-1}, u_{N-1}) + \mathcal{G}_N(x_N)\big\}\Big\}\,, \end{aligned} \tag{4.22}$$

where in the second equality we apply the fact that only terms $K(x_{N-1}, u_{N-1})$ and $\mathcal{G}_N(x_N)$ depend on u_{N-1}. Then by the assumption that $u_{N-1} = F_{N-1}x_{N-1}$ minimizes $\mathcal{G}_{N-1}(x_{N-1})$, it follows that

$$\begin{aligned} &\inf_{u_{N-1}\in\mathbb{L}^2}\big[K(x_{N-1}, u_{N-1}) + \mathcal{G}_N(x_N)\big] \\ &\quad = \inf_{u_{N-1}\in\mathbb{L}^2}\big[K(x_{N-1}, u_{N-1}) + \mathcal{G}_N(\mathcal{A}x_{N-1} + \tau\mathcal{A}_0\Pi_h u_{N-1})\big] \\ &\quad = K\big(x_{N-1}, F_{N-1}x_{N-1}\big) + \mathcal{G}_N\big(\mathcal{A}x_{N-1} + \tau\mathcal{A}_0\Pi_h(F_{N-1}x_{N-1})\big) \\ &\quad = \mathcal{G}_{N-1}(x_{N-1})\,. \end{aligned} \tag{4.23}$$

Now (4.22) and (4.23) lead to

$$\inf_{u.\in\mathbb{U}(t_l,T)} \mathcal{G}_{h,\tau}(t_l,z;u.) = \inf_{u_n\in\mathbb{L}^2, l\le n\le N-2} \Big[\sum_{n=l}^{N-2} K(x_n,u_n) + \mathcal{G}_{N-1}(x_{N-1})\Big]$$

$$= \inf_{u_n\in\mathbb{L}^2, l\le n\le N-3} \Big\{\sum_{n=l}^{N-3} K(x_n,u_n) + \inf_{u_{N-2}\in\mathbb{L}^2}\big[K(x_{N-2},u_{N-2}) + \mathcal{G}_{N-1}(x_{N-1})\big]\Big\},$$

which, and the assumption that $u_{N-2} = F_{N-2}x_{N-2}$ minimizes $\mathcal{G}_{N-2}(x_{N-2})$, yield

$$\inf_{u.\in\mathbb{U}(t_l,T)} \mathcal{G}_{h,\tau}(t_l,z;u.) = \inf_{u_n\in\mathbb{L}^2, l\le n\le N-3} \Big\{\sum_{n=l}^{N-3} K(x_n,u_n) + \mathcal{G}_{N-2}(x_{N-2})\Big\}.$$

By continuing this procedure we conclude that

$$\inf_{u.\in\mathbb{U}(t_l,T)} \mathcal{G}_{h,\tau}(t_l,z;u.) = \inf_{u_l\in\mathbb{L}^2}\big\{K(x_l,u_l) + \mathcal{G}_{l+1}(x_{l+1})\big\} = \mathcal{G}_l(x_l) = \mathcal{G}_l(z_0),$$

which is the assertion (4.21) and settles the claim.

(2) Claim: For any $x\in\mathbb{V}_h, l = 0,1,\ldots,N-1$ and any $n = l, l+1,\ldots,N-1$, $u = -\mathcal{K}_n^{-1}\mathcal{H}_n x$ minimizes $\mathcal{G}_n(x)$, and

$$\mathcal{G}_n(x) = \frac{1}{2}(\mathcal{P}_n x, x)_{\mathbb{L}^2},$$

where $\mathcal{K}.$, $\mathcal{H}.$ are defined in (4.20).

Indeed, for any $x\in\mathbb{V}_h$, it is obvious that $\mathcal{G}_N(x) = \frac{\alpha}{2}\|x\|^2 = \frac{1}{2}(\mathcal{P}_N x, x)_{\mathbb{L}^2}$. Then we proceed by induction and suppose that $\mathcal{G}_{n+1}(x) = \frac{1}{2}(\mathcal{P}_{n+1}x, x)_{\mathbb{L}^2}$. First notice that $K(x,u)\ge 0$. Hence $\mathcal{G}_{n+1}(x)\ge 0$, which means that $\mathcal{P}_{n+1}\in\mathbb{S}_+(\mathbb{L}^2|_{\mathbb{V}_h})$. Subsequently, we see that $\mathcal{K}_n = \mathbb{1}_h + \tau\mathcal{A}_0\mathcal{P}_{n+1}\mathcal{A}_0$ is invertible.

Based on the assumption, we have

$$\begin{aligned}
&K(x,u) + \mathcal{G}_{n+1}(\mathcal{A}_0 x + \tau\mathcal{A}_0\Pi_h u)\\
&\quad= \frac{1}{2}\Big[\tau\|x\|^2 + \tau\|u\|^2 + \Big(\mathcal{P}_{n+1}\big[\mathcal{A}_0 x + \tau\mathcal{A}_0\Pi_h u\big], \mathcal{A}_0 x + \tau\mathcal{A}_0\Pi_h u\Big)_{\mathbb{L}^2}\Big]\\
&\quad= \frac{1}{2}\Big[\big(\big[\tau\mathbb{1}_h + \mathcal{H}_n\big]x, x\big)_{\mathbb{L}^2} + \tau(\mathcal{K}_n u, u)_{\mathbb{L}^2} + 2\tau(\mathcal{H}_n x, u)_{\mathbb{L}^2}\Big].
\end{aligned}$$

By completing the squares, we obtain

$$\begin{aligned}
&K(x,u) + \mathcal{G}_{n+1}(\mathcal{A}_0 x + \tau\mathcal{A}_0\Pi_h u)\\
&\quad= \frac{1}{2}\Big[\tau\big(\mathcal{K}_n\big[u + \mathcal{K}_n^{-1}\mathcal{H}_n x\big], u + \mathcal{K}_n^{-1}\mathcal{H}_n x\big)_{\mathbb{L}^2}\\
&\qquad+ \big(\big[\tau\mathbb{1}_h + \mathcal{H}_n\big]x, x\big)_{\mathbb{L}^2} - \tau\big(\mathcal{H}_n\mathcal{K}_n^{-1}\mathcal{H}_n x, x\big)_{\mathbb{L}^2}\Big]\\
&\quad= \frac{1}{2}\Big[\tau\Big(\mathcal{K}_n\big[u + \mathcal{K}_n^{-1}\mathcal{H}_n x\big], u + \mathcal{K}_n^{-1}\mathcal{H}_n x\Big)_{\mathbb{L}^2} + (\mathcal{P}_n x, x)_{\mathbb{L}^2}\Big],
\end{aligned}$$

which yields

$$\mathcal{G}_n(x) = \inf_{u \in \mathbb{L}^2} \big\{ K(x,u) + \mathcal{G}_{n+1}(\mathcal{A}_0 x + \tau \mathcal{A}_0 \Pi_h u) \big\} = \frac{1}{2}(\mathcal{P}_n x, x)_{\mathbb{L}^2} .$$

That concludes the inductive proof.

(3) The steps (1) and (2) then complete the proof. □

Here we emphasize that $\mathcal{P}_l \in \mathbb{S}_+(\mathbb{L}^2|_{\mathbb{V}_h})$ for any $l = 0, 1, \ldots, N$; for the used notations, we again refer to Sect. 2.1. Also, (4.19) leads to

$$(\mathcal{P}_l \Pi_h z, z)_{\mathbb{L}^2} \leq 2 \mathcal{G}_{h,\tau}(t_l, z, u_\cdot) \qquad \forall z \in \mathbb{L}^2 \quad \forall u_\cdot \in \mathbb{U}(t_l, T) .$$

These properties make the difference Riccati equation (4.20) stand out if compared to standard discretization (see e.g. [10–12]), and allow to verify optimal convergence behavior for (4.20) in Sect. 4.2.

Remark 4.1 An essential step in the above program—which is detailed in Sect. 4.2—is to study Problem $(\mathbf{LQ})_{\mathtt{aux}}^{\cdot;h,\tau}$, i.e., the temporal discretization of Problem $(\mathbf{LQ})_{\mathtt{aux}}^{\cdot;h}$.

4.2 Approximation of the Riccati Equation

To obtain the convergence rates in different norms— $\mathcal{L}(\mathbb{L}^2)$ and $\mathcal{L}(\mathbb{H}_0^1; \mathbb{L}^2)$—for the discretization of Riccati equation (1.9), we divide this section into two parts: Sects. 4.2.1 and 4.2.2 focus on the first consistent discretization $\mathcal{P}_\cdot$ that is given by (4.20), and its modification $\widetilde{\mathcal{P}}_\cdot$ proposed by (4.57). The related error analysis requires assumption **(H)**, and is based on the close connection of $\mathcal{P}(\cdot)$ with Problem $(\mathbf{LQ})_{\mathtt{aux}}^{\cdot}$, and of $\mathcal{P}_\cdot$ with Problem $(\mathbf{LQ})_{\mathtt{aux}}^{\cdot;h,\tau}$. Note that the solution $\widetilde{\mathcal{P}}_\cdot$ to the modified difference Riccati equation (4.57) ceases to take values in $\mathbb{S}_+(\mathbb{L}^2|_{\mathbb{V}_h})$, but only in $\mathbb{S}(\mathbb{L}^2|_{\mathbb{V}_h})$; see Remark 4.2 for further details. However, this scheme drastically cuts down the computational complexity of (4.20), while the same convergence behavior is maintained. Section 4.2.3 presents the second consistent discretization $\bar{\mathcal{P}}_\cdot$ provided in (4.68), as well as its modification $\widehat{\mathcal{P}}_\cdot$ in (4.94); there results are derived without assumption **(H)**.

In the following, we use the notations $x\big(\cdot; t, z, u(\cdot)\big)$, $x_h\big(\cdot; t, z, u_h(\cdot)\big)$, and $x_\cdot(t_l, z, u_\cdot)$ to denote the solutions of (4.2), (4.11), and (4.16), respectively, in order to emphasize their dependence on the initial conditions and the associated control variables; when there is no danger of confusion, we also write $x(\cdot; t, z)$, $x_h(\cdot; t, z)$, $x_\cdot(t_l, z)$ for the sake of simplicity. Similarly, we denote by $u^*(\cdot; t, z)$, $u_h^*(\cdot; t, z)$, $u_\cdot^*(t_l, z)$ the optimal control of Problems $(\mathbf{LQ})_{\mathtt{aux}}^{t}$, $(\mathbf{LQ})_{\mathtt{aux}}^{t;h}$, and $(\mathbf{LQ})_{\mathtt{aux}}^{t_l;h,\tau}$, respectively. Let us recall that all state and control variables in this section are deterministic, which is different from related quantities used in Sect. 4.3; they are introduced here *only* to verify the main results (i.e., Theorems 4.2, 4.3, 4.5 and 4.6) which is a part of our scheme to approximate Problem **(SLQ)**; see steps (4) and (5) in Fig. 4.1.

4.2.1 Estimates for the Spatially Semi-Discrete Riccati Equation

In the following result, we establish that $\mathcal{P}_h(\cdot)$ is uniformly bounded, and that the optimal pair $\big(x_h^*(\cdot), u_h^*(\cdot)\big)$ of Problem $(\mathbf{LQ})^{t;h}_{\text{aux}}$ is bounded by the initial state.

Lemma 4.2 *Let $\mathcal{P}_h(\cdot)$ be the solution of Riccati equation (4.8) and for any $t \in [0, T)$, $z \in \mathbb{L}^2$, let $\big(x_h^*(\cdot; t, z), u_h^*(\cdot; t, z)\big)$ be the optimal pair of Problem $(\mathbf{LQ})^{t;h}_{\text{aux}}$. Then, there exists a constant C independent of h such that*

$$\sup_{t\in[0,T]} \|\mathcal{P}_h(t)\|_{\mathcal{L}(\mathbb{L}^2|_{\mathbb{V}_h})} \leq C\,, \tag{4.24}$$

$$\sup_{t\in[0,T]} \Big[\sup_{s\in[t,T]} \big[\|x_h^*(s; t, z)\|^2 + \|u_h^*(s; t, z)\|^2\big] + \int_t^T \|x_h^*(s; t, z)\|^2_{\mathbb{H}_0^1} + \|u_h^*(s; t, z)\|^2_{\mathbb{H}_0^1}\, ds \Big] \leq C\|z\|^2\,. \tag{4.25}$$

Proof (1) Verification of (4.24). This assertion has been derived in Lemma 3.6 by virtue of Problem $(\mathbf{SLQ})^{\cdot;h}_{\text{aux}}$; and with the same idea it can also be obtained by the explicit representation (4.13) for the value function $V_h(\cdot,\cdot)$. Indeed, for any $z \in \mathbb{L}^2$, by the fact that $\mathcal{P}_h(t) \in \mathbb{S}_+(\mathbb{L}^2|_{\mathbb{V}_h})$ and (4.13), it follows that

$$\begin{aligned}
0 &\leq \big(\mathcal{P}_h(t)\Pi_h z, \Pi_h z\big)_{\mathbb{L}^2} = \big(\mathcal{P}_h(t)\Pi_h z, z\big)_{\mathbb{L}^2} \leq 2\mathcal{G}_h(t, z; 0) \\
&= \frac{1}{2}\int_t^T \|x_h(s; t, z, 0)\|^2\, \mathrm{d}s + \frac{\alpha}{2}\|x_h(T; t, z, 0)\|^2 \\
&\leq C\|\Pi_h z\|^2\,,
\end{aligned}$$

where C is independent of t, which settles the assertion (4.24).

(2) Verification of (4.25). The feedback law (4.12) and PDE (4.11) imply that

$$\begin{aligned}
&\|x_h^*(s; t, z)\|^2 + 2\int_t^s \|\nabla x_h^*(\theta; t, z)\|^2\, \mathrm{d}\theta \\
&\quad\leq \|\Pi_h z\|^2 + 2\sup_{t\in[0,T]} \|\mathcal{P}_h(t)\|_{\mathcal{L}(\mathbb{L}^2|_{\mathbb{V}_h})} \int_t^s \|x_h^*(\theta; t, z)\|^2\, \mathrm{d}\theta\,,
\end{aligned}$$

which, together with assertion (4.24) and Gronwall's inequality, leads to

$$\sup_{t\in[0,T]} \Big[\sup_{s\in[t,T]} \|x_h^*(s; t, z)\|^2 + \int_t^T \|\nabla x_h^*(\theta; t, z)\|^2\, \mathrm{d}\theta \Big] \leq C\|\Pi_h z\|^2\,. \tag{4.26}$$

Then we adopt Pontryagin's maximum principle to verify the assertion on $u_h^*(\cdot; t, z)$. By the deterministic version of (1.5)–(1.6) or [6, Chap. 12], we know that

$$u_h^*(s; t, z) = y_h(s) \qquad t \in [0, T), s \in [t, T], \tag{4.27}$$

where $y_h(\cdot)$ solves

$$\begin{cases} y_h'(s) = -\mathcal{A}_h y_h(s) + x_h^*(s; t, z) & s \in [t, T), \\ y_h(T) = -\alpha x_h^*(T; t, z). \end{cases} \tag{4.28}$$

Relying on the stability property of $y_h(\cdot)$, (4.26) and (4.27), we have

$$\begin{aligned} &\sup_{s\in[t,T]} \|u_h^*(s; t, z)\|^2 + \int_t^T \|\nabla u_h^*(s; t, z)\|^2 \, \mathrm{d}s \\ &\quad \le C\Big[\|x_h^*(T; t, z)\|^2 + \int_t^T \|x_h^*(s; t, z)\|^2 \, \mathrm{d}s\Big] \\ &\quad \le C\|z\|^2, \end{aligned}$$

where C is independent of t. That completes the proof. □

In the same vein, we can prove the following result for the solution $\mathcal{P}(\cdot)$ to Riccati equation (1.9).

Lemma 4.3 *Fix $t \in [0, T]$ and $z \in \mathbb{L}^2$. Let $\big(x^*(\cdot; t, z), u^*(\cdot; t, z)\big)$ be the optimal pair of Problem* **(LQ)**$^t_{\mathrm{aux}}$. *Let $\mathcal{P}(\cdot)$ be the solution to Riccati equation* (1.9). *There exists C such that*

$$\sup_{t\in[0,T]} \|\mathcal{P}(t)\|_{\mathcal{L}(\mathbb{L}^2)} \le C, \tag{4.29}$$

$$\begin{aligned} &\sup_{t\in[0,T]} \Big[\sup_{s\in[t,T]} \big[\|x^*(s; t, z)\|^2 + \|u^*(s; t, z)\|^2\big] \\ &\quad + \int_t^T \|x^*(s; t, z)\|^2_{\mathbb{H}_0^1} + \|u^*(s; t, z)\|^2_{\mathbb{H}_0^1} \, ds\Big] \le C\|z\|^2. \end{aligned} \tag{4.30}$$

For any given $t \in [0, T)$, to estimate $\|\mathcal{P}(t) - \mathcal{P}_h(t)\Pi_h\|_{\mathcal{L}(\mathbb{L}^2)}$ we follow a related strategy as in [13, Sect. 3]: For $z \in \mathbb{L}^2$, we shall compare $\big(\mathcal{P}(t)z, z\big)_{\mathbb{L}^2}$ and $\big(\mathcal{P}_h(t)\Pi_h z, z\big)_{\mathbb{L}^2}$. There are two possibilities.

Case (i) $\big(\mathcal{P}_h(t)\Pi_h z, z\big)_{\mathbb{L}^2} \le \big(\mathcal{P}(t)z, z\big)_{\mathbb{L}^2}$.

In this case, based on the explicit representations of the value functions $V(\cdot, \cdot)$ resp. $V_h(\cdot, \cdot)$ in (4.5) resp. (4.13), we arrive at

$$\begin{aligned} 0 &\le \big(\mathcal{P}(t)z, z\big)_{\mathbb{L}^2} - \big(\mathcal{P}_h(t)\Pi_h z, z\big)_{\mathbb{L}^2} = 2V(t, z) - 2V_h(t, z) \\ &= 2\mathcal{G}\big(t, z; u^*(\cdot; t, z)\big) - 2\mathcal{G}_h\big(t, z; u_h^*(\cdot; t, z)\big). \end{aligned} \tag{4.31}$$

To estimate the difference of the right-hand side of (4.31), we replace $u^*(\cdot;t,z)$ by the (possibly) non-optimal control $u_h^*(\cdot;t,z) \in L^2(t,T;\mathbb{L}^2)$. By the optimality of $u^*(\cdot;t,z)$ in Problem $(\mathbf{LQ})^t_{\rm aux}$, we then conclude that

$$\begin{aligned} 0 &\le \big(\mathcal{P}(t)z, z\big)_{\mathbb{L}^2} - \big(\mathcal{P}_h(t)\Pi_h z, z\big)_{\mathbb{L}^2} \\ &\le 2\mathcal{G}\big(t,z;u_h^*(\cdot;t,z)\big) - 2\mathcal{G}_h\big(t,z;u_h^*(\cdot;t,z)\big)\,. \end{aligned} \tag{4.32}$$

Hence, to bound the difference of the quadratic cost functionals, we have to estimate $\big\|x\big(\cdot;t,z,u_h^*(\cdot;t,z)\big) - x_h^*\big(\cdot;t,z,u_h^*(\cdot;t,z)\big)\big\|$, which will be done in Lemma 4.4 below.

Case (ii) $\big(\mathcal{P}_h(t)\Pi_h z, z\big)_{\mathbb{L}^2} > \big(\mathcal{P}(t)z, z\big)_{\mathbb{L}^2}$.

In this case, similar to Case (i), we can deduce that

$$\begin{aligned} 0 &\le \big(\mathcal{P}_h(t)\Pi_h z, z\big)_{\mathbb{L}^2} - \big(\mathcal{P}(t)z, z\big)_{\mathbb{L}^2} \\ &\le 2\mathcal{G}_h\big(t,z;u^*(\cdot;t,z)\big) - 2\mathcal{G}\big(t,z;u^*(\cdot;t,z)\big)\,. \end{aligned} \tag{4.33}$$

Note that $u^*(\cdot;t,z) \in L^2(t,T;\mathbb{L}^2)$ may be a non-optimal control in Problem $(\mathbf{LQ})^{t;h}_{\rm aux}$. Then, the difference between $x_h\big(\cdot;t,z,u^*(\cdot;t,z)\big)$ and $x^*\big(\cdot;t,z,u^*(\cdot;t,z)\big)$ need be estimated, which will be dealt with in Lemma 4.5.

Lemma 4.4 *For any $t \in [0,T)$ and $z \in \mathbb{L}^2$, suppose that $\big(x_h^*(\cdot;t,z), u_h^*(\cdot;t,z)\big)$ is the optimal pair of Problem $(\mathbf{LQ})^{t;h}_{\rm aux}$. Let $\bar{x}(\cdot;t,z) \triangleq x\big(\cdot;t,z,u_h^*(\cdot;t,z)\big)$. Then there exists a constant C independent of h such that*

$$\sup_{t\in[0,T]}\sup_{s\in[t,T]} \|\bar{x}(s;t,z)\| \le C\|z\|\,, \tag{4.34}$$

$$\|\bar{x}(s;t,z) - x_h^*(s;t,z)\| \le Ch^2\Big[\frac{1}{s-t} + \ln\frac{1}{h}\Big]\|z\| \qquad \forall\, s \in (t,T]\,, \tag{4.35}$$

$$\int_t^T \|\bar{x}(s;t,z) - x_h^*(s;t,z)\|\,ds \le Ch^2\ln\frac{1}{h}\|z\|\,. \tag{4.36}$$

Proof Throughout the proof, we write $\bar{x}(\cdot)$ for $\bar{x}(\cdot;t,z)$. Recall that $\widehat{E}(\cdot)$ resp. $\widehat{E}_h(\cdot)$ denote semigroups generated by $\mathcal{A}$ resp. $\mathcal{A}_h$, and that $\widehat{G}(\cdot) = \widehat{E}(\cdot) - \widehat{E}_h(\cdot)\Pi_h$; see Sect. 4.1.2.

(1) Verification of (4.34). By (4.2), Lemmata 2.1 and 4.2, it follows that

$$\begin{aligned} \|\bar{x}(s)\| &= \Big\|\widehat{E}(s-t)z + \int_t^s \widehat{E}(s-\theta)u_h^*(\theta;t,z)\,\mathrm{d}\theta\Big\| \\ &\le C\Big[\|z\| + \sup_{s\in[t,T]}\|u_h^*(s;t,z)\|\Big] \\ &\le C\|z\|\,, \end{aligned}$$

where C is independent of h and t. That settles (4.34).

(2) Verification of (4.35). By (4.2), (4.11) and Lemmas 4.2, 2.3 for $\widehat{G}_h(\cdot)$ with $(\rho,\gamma)=(0,0)$ and $(\rho,\gamma)=(0,2)$, we find that[1]

$$\begin{aligned}
\|\bar{x}(s)-x_h^*(s;t,z)\| &\le \big\|\widehat{G}_h(s-t)z\big\| + \Big\| \int_{(s-h^2)\vee t}^{s} \widehat{G}_h(s-\theta)u_h^*(\theta;t,z)\,\mathrm{d}\theta \Big\| \\
&\quad + \Big\| \int_t^{(s-h^2)\vee t} \widehat{G}_h(s-\theta)u_h^*(\theta;t,z)\,\mathrm{d}\theta \Big\| \\
&\le C\frac{h^2}{s-t}\|z\| + Ch^2 \sup_{s\in[t,T]} \|u_h^*(s;t,z)\| \\
&\quad + C\int_t^{(s-h^2)\vee t} \frac{h^2}{s-\theta}\,\mathrm{d}\theta \sup_{s\in[t,T]} \|u_h^*(s;t,z)\| \\
&\le Ch^2\Big[\frac{1}{s-t} + 1 + \ln(s-t) - \ln h\Big]\|z\|,
\end{aligned}$$

which yields the assertion (4.35) since

$$\ln(s-t) - \ln h \le \ln T + \ln\frac{1}{h} = \ln\frac{1}{h}\Big[\frac{\ln T}{\ln\frac{1}{h_0}} + 1\Big] = \big[C_{h_0}+1\big]\ln\frac{1}{h}.$$

(3) Verification of (4.36). Based on systems (4.2), (4.11) and assertions (4.25), (4.34), (4.35), we arrive at

$$\begin{aligned}
&\int_t^T \|\bar{x}(s;t,z) - x_h^*(s;t,z)\|\,\mathrm{d}s \\
&\le \int_t^{(t+h^2)\wedge T} \sup_{s\in[t,T]} \big[\|\bar{x}(s;t,z)\| + \|x_h^*(s;t,z)\|\big]\,\mathrm{d}s \\
&\quad + Ch^2 \int_{(t+h^2)\wedge T}^{T} \Big[\frac{1}{s-t} + \ln\frac{1}{h}\Big]\|z\|\,\mathrm{d}s \\
&\le Ch^2 \ln\frac{1}{h}\|z\|.
\end{aligned}$$

That completes the proof. □

In the same vein as in Lemma 4.4, we derive the following

Lemma 4.5 *For any $t\in[0,T)$ and $z\in\mathbb{L}^2$, let $\big(x^*(\cdot;t_l,z),u^*(\cdot;t_l,z)\big)$ be the optimal pair of Problem* $(\mathbf{LQ})_{\mathrm{aux}}^{t_l}$. *Denote by $\widehat{x}_h(\cdot;t,z) \triangleq x_h(\cdot;t,z,u^*(\cdot;t,z))$. Then*

[1] Note that to estimate $\big\|\int_t^s \widehat{G}_h(s-\theta)u_h^*(\theta;t,z)\,\mathrm{d}\theta\big\|$, if we only apply Lemma 2.3 with $(\rho,\gamma)=(0,2)$, this would lead to the bound $\big\|\int_t^s \widehat{G}_h(s-\theta)u_h^*(\theta;t,z)\,\mathrm{d}\theta\big\| \le Ch^2\|z_l\|\int_t^s \frac{1}{s-\theta}\,\mathrm{d}\theta$, the right-hand side of which would be infinite. Hence, to deduce a rate we divide this integral into two parts: $\int_{(s-h^2)\vee t}^s \cdots\mathrm{d}\theta$ and $\int_t^{(s-h^2)\vee t}\cdots\mathrm{d}\theta$, and we then use Lemma 2.3 with $(\rho,\gamma)=(0,0)$ and $(\rho,\gamma)=(0,2)$ to estimate these two parts. This trick will be used frequently throughout this chapter.

$$\sup_{t\in[0,T]}\sup_{s\in[t,T]}\|\widehat{x}_h(s;t,z)\| \le C\|z\|\,,$$
$$\|\widehat{x}_h(s;t,z)-x^*(s;t,z)\| \le Ch^2\Big[\frac{1}{s-t}+\ln\frac{1}{h}\Big]\|z\| \qquad \forall\, s\in(t,T]\,,$$
$$\int_t^T \|\widehat{x}_h(s;t,z)-x^*(s;t,z)\|\,ds \le Ch^2\ln\frac{1}{h}\|z\|\,.$$

So far, we have obtained the following results in this section:

(a) By (4.29) in Lemma 4.3, the $\mathcal{L}(\mathbb{L}^2)$-valued solution $\mathcal{P}(\cdot)$ to the Riccati equation (1.9) is bounded; by its close connection to Problem $(\mathbf{LQ})^t_{\mathrm{aux}}$—see formulae (4.4) and (4.5)—the stability bound (4.29) for $\mathcal{P}(\cdot)$ leads to the bound (4.30) for the optimal pair $\big(x^*(\cdot;t,z),u^*(\cdot;t,z)\big)$ of Problem $(\mathbf{LQ})^t_{\mathrm{aux}}$.
(b) According to Lemma 4.2, these stability bounds are inherited by $\mathcal{P}_h(\cdot)$, which solves the spatial discretization (4.8). This family of $\mathcal{L}(\mathbb{L}^2|_{\mathbb{V}_h})$-valued operators is also linked to Problem $(\mathbf{LQ})^{t;h}_{\mathrm{aux}}$, whose optimal pairs $\big(x^*_h(\cdot;t,z),u^*_h(\cdot;t,z)\big)$ are bounded (uniformly in h) in terms of z; see (4.25).
(c) In Lemma 4.4, we compare the solution $\bar{x}(\cdot;t,z)$ to PDE (4.2) (with control $u(\cdot)=u^*_h(\cdot;t,z)$) with the optimal state $x^*_h(\cdot;t,z)$ of Problem $(\mathbf{LQ})^{t;h}_{\mathrm{aux}}$.
(d) The 'reverse case' is addressed in Lemma 4.5, where we compare the solution $\widehat{x}(\cdot;t,z)$ to semi-discretization (4.11) (with the given control $u_h(\cdot)=u^*(\cdot;t,z)$) with the optimal state $x^*(\cdot;t,z)$ of Problem $(\mathbf{LQ})^t_{\mathrm{aux}}$.

The following theorem is the main result in this section, which bounds the error between $\mathcal{P}(\cdot)$ and $\mathcal{P}_h(\cdot)$.

Theorem 4.1 *Suppose that $\mathcal{P}(\cdot)$ and $\mathcal{P}_h(\cdot)$ are solutions to Riccati equations* (4.7) *and* (4.8) *respectively. Then there exists a constant C independent of h such that*

$$\|\mathcal{P}(t)-\mathcal{P}_h(t)\Pi_h\|_{\mathcal{L}(\mathbb{L}^2)} \le Ch^2\Big[\frac{\alpha}{T-t}+\ln\frac{1}{h}\Big] \qquad \forall\, t\in[0,T)\,. \tag{4.37}$$

Proof For given $t\in[0,T)$, and $z\in\mathbb{L}^2$, based on (4.32) and (4.33) we split the proof into the following two cases.

Case (i). $(\mathcal{P}_h(t)\Pi_h z,z)_{\mathbb{L}^2} \le (\mathcal{P}(t)z,z)_{\mathbb{L}^2}$.

In this case, by (4.32) and the definition of the cost functionals $\mathcal{G}(\cdot,\cdot;\cdot)$ and $\mathcal{G}_h(\cdot,\cdot;\cdot)$, as well as $\bar{x}(\cdot;t,z)$ that is defined in Lemma 4.4, we may use binomial formula and estimate

$$\begin{aligned}
0 &\le \big(\mathcal{P}(t)z,z\big)_{\mathbb{L}^2}-\big(\mathcal{P}_h(t)\Pi_h z,z\big)_{\mathbb{L}^2} \le 2\mathcal{G}\big(t,z;u^*_h(\cdot;t,z)\big)-2\mathcal{G}_h\big(t,z;u^*_h(\cdot;t,z)\big)\\
&= \int_t^T \big[\|\bar{x}(s;t,z)\|^2-\|x^*_h(s;t,z)\|^2\big]\,\mathrm{d}s+\alpha\big[\|\bar{x}(T;t,z)\|^2-\|x^*_h(T;t,z)\|^2\big]\\
&\le C\sup_{s\in[t,T]}\big[\|\bar{x}(s;t,z)\|+\|x^*_h(s;t,z)\|\big]\\
&\quad\times\Big[\int_t^T \|\bar{x}(s;t,z)-x^*_h(s;t,z)\|\,\mathrm{d}s+\alpha\|\bar{x}(T;t,z)-x^*_h(T;t,z)\|\Big]
\end{aligned}$$

$$\leq Ch^2\Big[\frac{\alpha}{T-t} + \ln\frac{1}{h}\Big]\|z\|^2 .$$

Case (ii). $\big(\mathcal{P}_h(t)\Pi_h z, z\big)_{\mathbb{L}^2} > \big(\mathcal{P}(t)z, z\big)_{\mathbb{L}^2}$.

As in Case (i), by (4.33) and Lemmata 4.3 and 4.5, we can deduce that

$$\begin{aligned}
0 &\leq \big(\mathcal{P}_h(t)\Pi_h z, z\big)_{\mathbb{L}^2} - \big(\mathcal{P}(t)z, z\big)_{\mathbb{L}^2} \leq 2\mathcal{G}_h(t, z; u^*(\cdot; t, z)) - 2\mathcal{G}(t, z; u^*(\cdot; t, z)) \\
&\leq C \sup_{s\in[t,T]} \big[\|\widehat{x}_h(s; t, z)\| + \|x^*(s; t, z)\|\big] \\
&\quad \times \Big[\int_t^T \|\widehat{x}_h(s; t, z) - x^*(s; t, z)\| \,\mathrm{d}s + \alpha\|\widehat{x}_h(T; t, z) - x^*(T; t, z)\|\Big] \\
&\leq Ch^2\Big[\frac{\alpha}{T-t} + \ln\frac{1}{h}\Big]\|z\|^2 .
\end{aligned}$$

A combination of these two estimates leads to

$$\big|\big(\mathcal{P}_h(t)\Pi_h z, z\big)_{\mathbb{L}^2} - \big(\mathcal{P}(t)z, z\big)_{\mathbb{L}^2}\big| \leq Ch^2\Big[\frac{\alpha}{T-t} + \ln\frac{1}{h}\Big]\|z\|^2 ,$$

which implies the assertion (4.37). □

4.2.2 *The First Consistent Difference Riccati Equation*

The main result in this section is Theorem 4.2, which provides an error estimate for (4.20)—the temporal discretization of (4.8). For its verification, we parallel the steps in Sect. 4.2.1, and correspondingly introduce related quantities $\bar{x}_h(\cdot; t_l, z)$ in Lemma 4.8, and $\widehat{x}_\cdot(t_l, z)$ in Lemma 4.9 to quantify the temporal error inherent to (4.20) step by step.

The iterates of the *difference Riccati equation* (4.20) are known to take values in $\mathbb{S}_+(\mathbb{L}^2|_{\mathbb{V}_h})$; see Lemma 4.1. In this respect, this discretization is *consistent*, and will be studied in part **a**. of this section. In order to simplify (4.20), we later give up consistency to arrive at its modification (4.57) in part **b.**: the benefit next to its reduced complexity, however, is that its iterates still converge with the same order—a result that will be established by induction through a perturbation argument that starts with the *difference Riccati equation* (4.20).

a. The difference Riccati equation (4.20). The following result characterizes the solution of (4.16); since the proof is obvious, we skip it here.

Lemma 4.6 *For any* $l = 0, 1, \ldots, N-1$, $z \in \mathbb{L}^2$, $u_\cdot \in \mathbb{U}(t_l, T)$, *let* $x_\cdot(t_l, z, u_\cdot)$ *solve* (4.16). *Then for any* $n = l, l+1, \ldots, N$, *it holds that*

$$x_n(t_l, z, u_\cdot) = \mathcal{A}_0^{n-l}\Pi_h z + \tau\sum_{k=l}^{n-1}\mathcal{A}_0^{n-k}\Pi_h u_k . \tag{4.38}$$

The following lemma establishes that $\mathcal{P}_\cdot$ from (4.20) is uniformly bounded, and also provides a uniform bound for $x^*_\cdot(t_l, z)$ from Problem $(\mathbf{LQ})^{t_l;h,\tau}_{\mathrm{aux}}$, whose derivation is based on (4.38). It will turn out later that (4.38) is useful at other places as well, including the estimation of the difference between $x\big(\cdot; t_l, z, u(\cdot)\big)$ and $x_\cdot(t_l, z, u_\cdot)$; see Lemmata 4.8 and 4.9.

Lemma 4.7 *Let $\mathcal{P}_\cdot$ denote the solution to the difference Riccati equation* (4.20). *For any $l = 0, 1, \ldots, N-1$ and $z \in \mathbb{L}^2$, let $\big(x^*_\cdot(t_l, z), u^*_\cdot(t_l, z)\big)$ be the optimal pair of Problem $(\mathbf{LQ})^{t_l;h,\tau}_{\mathrm{aux}}$. Then, there exists a constant C independent of h and τ such that*

$$\max_{0\le l\le N-1} \|\mathcal{P}_l\|_{\mathcal{L}(\mathbb{L}^2|_{\mathbb{V}_h})} \le C\,, \tag{4.39}$$

$$\max_{0\le l\le N-1}\max_{l\le n\le N-1} \big[\|x^*_n(t_l, z)\| + \|u^*_n(t_l, z)\|\big] \le C\|z\|\,. \tag{4.40}$$

Proof (1) Verification of (4.39). This assertion can be obtained by the explicit representation (4.19) for the value function $V_{h,\tau}(\cdot,\cdot)$, and the fact that $\|\mathcal{A}_0\|_{\mathcal{L}(\mathbb{L}^2|_{\mathbb{V}_h})} \le 1$; indeed, for any $z \in \mathbb{L}^2$, by the fact that $\mathcal{P}_l \in \mathbb{S}_+(\mathbb{L}^2|_{\mathbb{V}_h})$, (4.19) and Lemma 4.6, it follows that

$$\begin{aligned}
0 &\le (\mathcal{P}_l \Pi_h z, z)_{\mathbb{L}^2} \le 2\mathcal{G}_{h,\tau}(t_l, z; 0)\\
&= \tau \sum_{n=l}^{N-1} \|\mathcal{A}_0^{n-l}\Pi_h z\|^2 + \alpha\|\mathcal{A}_0^{N-l}\Pi_h z\|^2\\
&\le (T+\alpha)\|\Pi_h z\|^2\,.
\end{aligned}$$

(2) Verification of (4.40). By the feedback law (4.18) and the difference equation (4.16), for any $n \ge l$, we have

$$x^*_{n+1}(t_l, z) = \mathcal{A}_0 x^*_n(t_l, z) - \tau\mathcal{A}_0\mathcal{K}_n^{-1}\mathcal{H}_n x^*_n(t_l, z)\,.$$

By the definition of $\mathcal{K}_\cdot$ and $\mathcal{H}_\cdot$ in (4.20), assertion (4.39) and the fact that $\{\mathcal{P}_n\}_{n=0}^N \subset \mathbb{S}_+(\mathbb{L}^2|_{\mathbb{V}_h})$, we arrive at

$$\|\mathcal{A}_0\mathcal{K}_n^{-1}\mathcal{H}_n\|_{\mathcal{L}(\mathbb{L}^2|_{\mathbb{V}_h})} \le \|\mathcal{P}_{n+1}\|_{\mathcal{L}(\mathbb{L}^2|_{\mathbb{V}_h})} \le C\,,$$

which leads to

$$\|x^*_n(t_l, z)\| \le (1 + C\tau)\|x^*_{n-1}(t_l, z)\| \le \cdots \le (1 + C\tau)^{n-l}\|\Pi_h z\| \le C\|\Pi_h z\|\,.$$

Finally, relying on the discrete feedback form (4.18) again, we have

$$\|u^*_n(t_l, z)\| = \|-\mathcal{K}_n^{-1}\mathcal{H}_n x^*_n(t_l, z)\| \le C\|x^*_n(t_l, z)\| \le C\|\Pi_h z\|\,.$$

That completes the proof. ☐

In the next step, we estimate $\|\mathcal{P}_h(t_l)\Pi_h - \mathcal{P}_l\Pi_h\|_{\mathcal{L}^2(\mathbb{L}^2)}$ by following arguments similar to those used in Sect. 4.2.1. The following two results (Lemmata 4.8 and 4.9) show some estimates in time, where similar estimates in space are presented in Lemmata 4.4 and 4.5.

Lemma 4.8 *For any $l = 0, 1, \ldots, N-1$ and $z \in \mathbb{L}^2$, let $\big(x_\cdot^*(t_l, z), u_\cdot^*(t_l, z)\big)$ be the optimal pair of Problem* $(\mathbf{LQ})_{\mathrm{aux}}^{t_l;h,\tau}$. *Suppose that*

$$\bar{u}_h(t; t_l, z) \triangleq u_n^*(t_l, z) \qquad \forall\, t \in [t_n, t_{n+1}), \quad n = l, l+1, \ldots, N-1, \tag{4.41}$$

and $\bar{x}_h(\cdot; t_l, z) \triangleq x_h\big(\cdot; t_l, z, \bar{u}(\cdot; t_l, z)\big)$. There exists a constant C independent of h, τ such that

$$\max_{0\le l\le N-1} \sup_{t\in[t_l,T]} \|\bar{x}_h(t; t_l, z)\| \le C\|z\|, \tag{4.42}$$

$$\sum_{n=l}^{N-1} \int_{t_n}^{t_{n+1}} \big[\|\bar{x}_h(t; t_l, z) - \bar{x}_h(t_n; t_l, z)\| + \|\bar{x}_h(t; t_l, z) - \bar{x}_h(t_{n+1}; t_l, z)\|\big]\, dt$$
$$\le C\tau \ln \frac{1}{\tau}\|z\|, \tag{4.43}$$

$$\|\bar{x}_h(t_n; t_l, z) - x_n^*(t_l, z)\| \le C\tau\Big[\frac{1}{t_n - t_l} + \ln\frac{1}{\tau}\Big]\|z\| \quad n = l+1, \cdots, N. \tag{4.44}$$

Proof For simplicity, we write $\bar{x}_h(\cdot)$ instead of $\bar{x}_h(\cdot; t_l, z)$ in the proof.

(1) Verification (4.42). By (4.11) and (4.40) in Lemma 4.7, it follows that

$$\begin{aligned}\|\bar{x}_h(t)\| &= \Big\|\widehat{E}_h(t - t_l)\Pi_h z + \int_{t_l}^{t} \widehat{E}_h(t-\theta)\bar{u}_h(\theta; t_l, z)\, \mathrm{d}\theta\Big\| \\ &\le C\big[\|\Pi_h z\| + \max_{l\le k\le N-1} \|u_k^*(t_l, z)\|\big] \\ &\le C\|\Pi_h z\|.\end{aligned}$$

(2) Verification of (4.43). We only prove

$$I := \sum_{n=l}^{N-1} \int_{t_n}^{t_{n+1}} \|\bar{x}_h(t) - \bar{x}_h(t_n)\|\, \mathrm{d}t \le C\tau \ln\frac{1}{\tau}\|z\|,$$

since the remaining part can be shown similarly. By (4.11), we find that

$$\begin{aligned} I \le& \sum_{n=l}^{N-1} \int_{t_n}^{t_{n+1}} \big\|\big[\widehat{E}_h(t - t_l) - \widehat{E}_h(t_n - t_l)\big]\Pi_h z\big\|\, \mathrm{d}t \\ &+ \sum_{n=l}^{N-1} \int_{t_n}^{t_{n+1}} \Big\|\int_{t_l}^{t_n} \big[\widehat{E}_h(t-s) - \widehat{E}_h(t_n - s)\big]\bar{u}_h(s; t_l, z)\, \mathrm{d}s\Big\|\, \mathrm{d}t\end{aligned}$$

$$
\begin{aligned}
&+\sum_{n=l}^{N-1}\int_{t_n}^{t_{n+1}}\left\|\int_{t_n}^{t}\widehat{E}_h(t-s)u_n^*(t_l,z)\,\mathrm{d}s\right\|\mathrm{d}t\\
&=: I_1+I_2+I_3\,. \qquad (4.45)
\end{aligned}
$$

Now we estimate the summands on the right-hand side of (4.45) independently. For I_1, by virtue of $\nu(\cdot)$ which is defined in (2.8), as well as (2.2a) for $\widehat{E}_h(\cdot)$ with $\gamma=0$, and (2.2c) for $\widehat{E}_h(\cdot)$ with $\gamma=1$ in Lemma 2.1, it follows that

$$
\begin{aligned}
I_1 &\le \int_{t_l}^{(t_l+2\tau)\wedge T} C\|z\|\,\mathrm{d}t+\int_{(t_l+2\tau)\wedge T}^{T} C\frac{t-\nu(t)}{\nu(t)-t_l}\|z\|\,\mathrm{d}t\\
&\le C\|z\|\Big[\tau+\tau\int_{(t_l+2\tau)\wedge T}^{T}\frac{1}{t-\tau-t_l}\,\mathrm{d}t\Big]\\
&\le C\|z\|\tau\ln\frac{1}{\tau}\,. \qquad (4.46)
\end{aligned}
$$

Here, the constant C comes from Lemma 2.1 and also depends on $C_{\tau_0}=\frac{\ln T}{\ln\frac{1}{\tau_0}}$; in the second inequality, to bound the integral (after computation) we apply the fact that $\frac{1}{\nu(t)-t_l}\le\frac{1}{t-\tau-t_l}$.

For I_2, we proceed accordingly to (4.46), and use (2.2c) in Lemma 2.1 to get

$$
\begin{aligned}
I_2 &\le C\sum_{n=l+1}^{N-1}\int_{t_n}^{t_{n+1}}\int_{t_n-\tau}^{t_n}\|\bar{u}(s;t_l,z)\|\,\mathrm{d}s\,\mathrm{d}t\\
&\quad +C\sum_{n=l+1}^{N-1}\int_{t_n}^{t_{n+1}}\int_{t_l}^{t_n-\tau}\frac{t-t_n}{t_n-s}\|\bar{u}(s;t_l,z)\|\,\mathrm{d}s\,\mathrm{d}t\\
&\le C\|z\|\Big[\tau+\tau\sum_{n=l}^{N-1}\int_{t_n}^{t_{n+1}}\int_{t_l}^{t_n-\tau}\frac{1}{t_n-s}\,\mathrm{d}s\,\mathrm{d}t\Big]\\
&\le C\|z\|\tau\ln\frac{1}{\tau}\,. \qquad (4.47)
\end{aligned}
$$

For I_3, by (4.40), it is easy to see that

$$
I_3\le C\tau\max_{l\le n\le N-1}\|u_n^*(t_l,z)\|\le C\tau\|z\|\,. \qquad (4.48)
$$

A combination of (4.45)–(4.48) now yields the assertion.

(3) Verification of (4.44). By (4.11) and (4.38), we find that

$$
\begin{aligned}
\|\bar{x}_h(t_n)-x_n^*(t_l,z)\| &\le \left\|\widehat{E}_h(t_n-t_l)\Pi_h z-\mathcal{A}_0^{n-l}\Pi_h z\right\|\\
&\quad+\left\|\sum_{j=l}^{n-1}\int_{(t_j,t_{j+1}]}\big[\widehat{E}_h(t_n-t)\Pi_h-\mathcal{A}_0^{n-j}\Pi_h\big]u_j^*(t_l,z)\,\mathrm{d}t\right\|\\
&=: J_1+J_2\,.
\end{aligned}
$$

We estimate these two terms independently. By (2.10c) in Lemma 2.4 with $\gamma = 2$ and (2.2c) in Lemma 2.1 with $\gamma = 1$, for any $\varepsilon \in (0, \tau)$ we have

$$\begin{aligned} J_1 &\leq \big\| \big[\widehat{E}_h(\varepsilon) - \mathbb{1}_h\big]\widehat{E}_h(t_n - t_l - \varepsilon)\Pi_h z\big\| + \big\|\widehat{G}_\tau(t_n - t_l - \varepsilon)z\big\| \\ &\leq C\frac{\varepsilon}{t_n - t_l - \varepsilon}\|z\| + C\tau\frac{1}{t_n - t_l - \varepsilon}\|z\|\,, \end{aligned}$$

where $C > 0$ is independent of ε. For the definition and properties of $\widehat{G}_\tau(\cdot)$, recall Sect. 4.1.2 and Lemma 2.4. Note that $\widehat{E}_h(t_n - t_l)z - \mathcal{A}_0^{n-l}\Pi_h z \neq \widehat{G}_\tau(t_n - t_l)z$. Hence, we divide it into two terms

$$\big[\widehat{E}_h(t_n - t_l) - \widehat{E}_h(t_n - t_l - \varepsilon)\big]\Pi_h z\,, \qquad \widehat{E}_h(t_n - t_l - \varepsilon)\Pi_h z - \mathcal{A}_0^{n-l}\Pi_h z\,,$$

and the latter one is just $\widehat{G}_\tau(t_n - t_l - \varepsilon)z$. Since $\varepsilon > 0$ is arbitrary, we conclude that

$$J_1 \leq C\tau\frac{1}{t_n - t_l}\|z\|\,.$$

Also, by (4.40) in Lemma 4.7 and (2.10c) in Lemma 2.4,

$$\begin{aligned} J_2 &\leq \sum_{j=l}^{n-1}\int_{(t_j,t_{j+1}]}\|\widehat{G}_\tau(t_n - t)\|_{\mathcal{L}(\mathbb{L}^2)}\max_{l\leq j\leq N-1}\|u_j^*(t_l, z)\|\,\mathrm{d}t \\ &\leq C\|z\|\Big[\int_{t_l}^{t_n-\tau}\frac{\tau}{t_n - t}\,\mathrm{d}t + \int_{t_n-\tau}^{t_n} C\,\mathrm{d}t\Big] \\ &\leq C\|z\|\tau\ln\frac{1}{\tau}\,. \end{aligned}$$

A combination of these two estimates now settles the assertion. □

Lemma 4.9 *For any $l = 0, 1, \ldots, N-1$ and $z \in \mathbb{L}^2$, let $\big(x_h^*(\cdot; t_l, z), u_h^*(\cdot; t_l, z)\big)$ be the optimal pair of Problem* $\textbf{(LQ)}_{\text{aux}}^{t_l;h}$. *Let*

$$\widehat{u}_n(t_l, z) \triangleq \frac{1}{\tau}\int_{t_n}^{t_{n+1}} u_h^*(t; t_l, z)\,dt \qquad \forall\, n = l, l+1, \ldots, N-1\,, \tag{4.49}$$

and $\widehat{x}_\cdot(t_l, z) \triangleq x_\cdot\big(t_l, z, \widehat{u}_\cdot(t_l, z)\big)$. Then there exists C independent of h and τ such that

$$\max_{0\leq l\leq N-1}\max_{l\leq n\leq N}\|\widehat{x}_n(t_l, z)\| \leq C\|z\|\,, \tag{4.50}$$

$$\begin{aligned} &\sum_{n=l}^{N-1}\int_{t_n}^{t_{n+1}}\big[\|x_h^*(t; t_l, z) - x_h^*(t_n; t_l, z)\| + \|x_h^*(t; t_l, z) - x_h^*(t_{n+1}; t_l, z)\|\big]\,dt \\ &\qquad \leq C\tau\ln\frac{1}{\tau}\|z\|\,, \end{aligned} \tag{4.51}$$

$$\|\widehat{x}_n(t_l, z) - x_h^*(t_n; t_l, z)\| \leq C\tau\Big[\frac{1}{t_n - t_l} + \ln\frac{1}{\tau}\Big]\|z\| \quad n = l+1, \cdots, N. \tag{4.52}$$

Proof (1) Verification of (4.50). By Lemmata 4.6 and 4.2, for any $n = l, l+1, \ldots, N$, we have

$$\begin{aligned}
\|\widehat{x}_n(t_l, z)\| &= \Big\|\mathcal{A}_0^{n-l}\Pi_h z + \sum_{k=l}^{n-1}\mathcal{A}_0^{n-k}\Pi_h \int_{t_k}^{t_{k+1}} u_h^*(t; t_l, z)\,\mathrm{d}t\Big\| \\
&\leq C\|\Pi_h z\| + C \sup_{t\in[t_l, T]} \|u_h^*(t; t_l, z)\| \\
&\leq C\|z\|.
\end{aligned}$$

(2) Verification of (4.51) and (4.52). As done for (4.43) and (4.44). □

The following lemma gives a regularity result for the optimal control to Problem $(\mathbf{LQ})_{\text{aux}}^{t_l}$, which will be needed below. Its derivation uses a tool from the 'open-loop approach'—but now for *deterministic* control problems.

Lemma 4.10 *For a given uniform partition I_τ of size $\tau \in (0, \tau_0] \subset (0, 1)$, for any $l = 0, 1, \ldots, N-1$ and $z \in \mathbb{L}^2$, suppose that $u_h^*(\cdot; t_l, z)$ is the optimal control of Problem $(\mathbf{LQ})_{\text{aux}}^{t_l;h}$. Then, for $s, s_0 \in [t_n, t_{n+1})$, $n = l, l+1, \ldots, N-1$, there exists a constant C independent of τ such that*

$$\|u_h^*(s; t_l, z) - u_h^*(s_0; t_l, z)\| \leq C\tau\Big[\frac{\alpha}{T - s\vee s_0} + \ln\frac{1}{\tau}\Big]\|z\|, \tag{4.53}$$

where $s \vee s_0 = \max\{s, s_0\}$ defined in (2.1).

Proof We adopt Pontryagin's maximum principle to verify the assertion. By the adjoint equation (4.28), it is easy to write its mild solution as

$$y_h(s; t_l, z) = -\alpha\widehat{E}_h(T-s)x_h^*(T; t_l, z) + \int_s^T \widehat{E}_h(\theta - s)x_h^*(\theta; t_l, z)\,\mathrm{d}\theta \quad s \in [t_l, T].$$

Then, with the help of (4.27), we arrive at (without loss of generality, we take $s_0 \leq s$)

$$\begin{aligned}
&\|u_h^*(s; t_l, z) - u_h^*(s_0; t_l, z)\| = \|y_h(s; t_l, z) - y_h(s_0; t_l, z)\| \\
&\leq \alpha\big\|\big[\widehat{E}_h(T-s) - \widehat{E}_h(T-s_0)\big]x_h^*(T; t_l, z)\big\| \\
&\quad + \Big\|\int_s^T \big[\widehat{E}_h(\theta - s_0) - \widehat{E}_h(\theta - s)\big]x_h^*(\theta; t_l, z)\,\mathrm{d}\theta\Big\| \\
&\quad + \Big\|\int_{s_0}^s \widehat{E}_h(\theta - s_0)x_h^*(\theta; t_l, z)\,\mathrm{d}\theta\Big\| \\
&=: \sum_{i=1}^3 I_i.
\end{aligned}$$

Lemmata 4.2 and 2.4 lead to

$$I_1 \leq C\tau \frac{\alpha}{T - s \vee s_0} \|z\| .$$

For I_2 , I_3, we can proceed similarly to deduce that

$$\begin{aligned} I_2 &\leq C \int_s^{(s+\tau)\wedge T} \sup_{t\in[t_l,T]} \|x_h^*(t; t_l, z)\| \, \mathrm{d}\theta \\ &\quad + C \int_{(s+\tau)\wedge T}^{T} \frac{s - s_0}{\theta - s} \sup_{t\in[t_l,T]} \|x_h^*(t; t_l, z)\| \, \mathrm{d}\theta \\ &\leq C\tau \ln \frac{1}{\tau} \|z\| , \end{aligned}$$

and

$$I_3 \leq C(s - s_0) \sup_{t\in[t_l,T]} \|x_h^*(t; t_l, z)\| \leq C\tau \|z\| .$$

Combining these estimates settles the assertion (4.53). □

The following result is comparable to Theorem 4.1, which was for $\mathcal{P}_h(\cdot)$ from the semi-discrete Riccati equation (4.8): now we address $\mathcal{P}_\cdot$ from the *difference Riccati equation* (4.20). Its proof is based on the results that we obtain so far in this section.

Theorem 4.2 *Let* $\mathcal{P}_h(\cdot)$ *resp.* $\mathcal{P}_\cdot$ *be solutions to Riccati equations* (4.8) *resp.* (4.20). *Then there exists a constant* C *independent of* h *and* τ *such that*

$$\|\mathcal{P}_h(t_l)\Pi_h - \mathcal{P}_l\Pi_h\|_{\mathcal{L}(\mathbb{L}^2)} \leq C\tau\Big[\frac{\alpha}{T - t_l} + \ln \frac{1}{\tau}\Big] \quad \forall l = 0, 1, \ldots, N - 1 . \tag{4.54}$$

Proof For given $l = 0, 1, \ldots, N - 1$, and $z \in \mathbb{L}^2$, we divide the proof into the following two cases.

Case (i). $(\mathcal{P}_l\Pi_h z, z)_{\mathbb{L}^2} \leq (\mathcal{P}_h(t_l)\Pi_h z, z)_{\mathbb{L}^2}$.

In this case, by (4.13), (4.19), and the definition of the cost functionals $\mathcal{G}_h(\cdot, \cdot; \cdot)$ and $\mathcal{G}_{h,\tau}(\cdot, \cdot; \cdot)$, as well as of $\bar{u}_h(\cdot; t_l, z)$ and $\bar{x}_h(\cdot; t_l, z)$ that are defined in Lemma 4.8, we can deduce that

$$\begin{aligned} 0 &\leq \big(\mathcal{P}_h(t_l)\Pi_h z, z\big)_{\mathbb{L}^2} - (\mathcal{P}_l\Pi_h z, z)_{\mathbb{L}^2} \\ &\leq 2\mathcal{G}_h\big(t_l, z; \bar{u}_h(\cdot; t_l, z)\big) - 2\mathcal{G}_{h,\tau}\big(t_l, z; u_\cdot^*(t_l, z)\big) \\ &= \sum_{n=l}^{N-1} \int_{t_n}^{t_{n+1}} \big[\|\bar{x}_h(t; t_l, z)\|^2 - \|\bar{x}_h(t_n; t_l, z)\|^2\big] \, \mathrm{d}t \\ &\quad + \sum_{n=l}^{N-1} \int_{t_n}^{t_{n+1}} \big[\|\bar{x}_h(t_n; t_l, z)\|^2 - \|x_n^*(t_l, z)\|^2\big] \, \mathrm{d}t \\ &\quad + \alpha\big[\|\bar{x}_h(T; t_l, z)\|^2 - \|x_N^*(t_l, z)\|^2\big] \end{aligned}$$

$$=: \sum_{i=1}^{3} I_i \,.$$

Here we apply the fact that $\|\bar{u}_h(\cdot; t_l, z)\|_{L^2(t_l,T;\mathbb{L}^2)} = \|u^*_\cdot(t_l, z)\|_{\mathbb{U}(t_l,T)}$ (see (4.41)). In what follows, we estimate the above three terms one by one.

For I_1, by (4.42) and (4.43) in Lemma 4.8, we have

$$\begin{aligned}
I_1 &\le \sum_{n=l}^{N-1} \int_{t_n}^{t_{n+1}} \big[\|\bar{x}_h(t; t_l, z)\| + \|\bar{x}_h(t_n; t_l, z)\|\big]\big[\|\bar{x}_h(t; t_l, z) - \bar{x}_h(t_n; t_l, z)\|\big]\, dt \\
&\le C\|z\| \sum_{n=l}^{N-1} \int_{t_n}^{t_{n+1}} \big[\|\bar{x}(t; t_l, z) - \bar{x}(t_n; t_l, z)\|\big]\, dt \\
&\le C\|z\|^2 \tau \ln \frac{1}{\tau} \,.
\end{aligned}$$

For I_2, by virtue of $\nu(\cdot)$, $\pi(\cdot)$ which are defined in (2.8), as well as Lemmata 4.7 and 4.8 it follows that

$$\begin{aligned}
I_2 &\le \int_{t_l}^{T} \big[\big\|\bar{x}_h\big(\nu(t); t_l, z\big)\big\| + \big\|x^*_{\pi(t)-1}(t_l, z)\big\|\big]\big[\big\|\bar{x}_h\big(\nu(t); t_l, z\big) - x^*_{\pi(t)-1}(t_l, z)\big\|\big]\, dt \\
&\le C\|z\| \int_{(t_{l+2})\wedge T}^{T} \big[\big\|\bar{x}_h\big(\nu(t); t_l, z\big) - x^*_{\pi(t)-1}(t_l, z)\big\|\big]\, dt + C\|z\|^2 \int_{t_l}^{(t_{l+2})\wedge T} 1\, dt \\
&\le C\tau\|z\|^2 + C\|z\|^2 \int_{(t_{l+2})\wedge T}^{T} \Big[\frac{\alpha}{\nu(t) - t_l} + \ln \frac{1}{\tau}\Big]\, dt \\
&\le C\tau \ln \frac{1}{\tau} \|z\|^2 \,.
\end{aligned}$$

For I_3, by Lemma 4.8, it is obvious that

$$I_3 \le C\tau\Big[\frac{\alpha}{T - t_l} + \ln \frac{1}{\tau}\Big]\|z\|^2 \,.$$

A combination of the above estimates now leads to

$$0 \le \big(\mathcal{P}(t_l)z, z\big)_{\mathbb{L}^2} - \big(\mathcal{P}_l \Pi_h z, z\big)_{\mathbb{L}^2} \le C\tau\Big[\frac{\alpha}{T - t_l} + \ln \frac{1}{\tau}\Big]\|z\|^2 \,. \tag{4.55}$$

Case (ii). $(\mathcal{P}_l \Pi_h z, z)_{\mathbb{L}^2} > (\mathcal{P}_h(t_l)\Pi_h z, z)_{\mathbb{L}^2}$.

In a similar way to Case (i), by Lemmata 4.2 and 4.9 we infer that

$$0 \le \big(\mathcal{P}_l \Pi_h z, z\big)_{\mathbb{L}^2} - \big(\mathcal{P}_h(t_l)\Pi_h z, z\big)_{\mathbb{L}^2}$$

$$
\begin{aligned}
&\leq C\|z\| \sum_{n=l}^{N-1} \int_{t_n}^{t_{n+1}} \|\widehat{x}_n(t_l, z) - x_h^*(t; t_l, z)\| \,\mathrm{d}t \\
&+C\|z\| \sum_{n=l}^{N-1} \int_{t_n}^{t_{n+1}} \left\| \frac{1}{\tau} \int_{t_n}^{t_{n+1}} u_h^*(\theta; t_l, z) \,\mathrm{d}\theta - u_h^*(t; t_l, z) \right\| \mathrm{d}t \\
&+C\|z\| \|\widehat{x}_N(t_l, z) - x_h^*(T; t_l, z)\| \\
&=: C\|z\| \sum_{i=1}^{3} J_i \,.
\end{aligned}
$$

For J_1, the triangle inequality yields

$$
J_1 \leq \sum_{n=l}^{N-1} \int_{t_n}^{t_{n+1}} \|\widehat{x}_n(t_l, z) - x_h^*(t_n; t_l, z)\| + \|x_h^*(t_n; t_l, z) - x_h^*(t; t_l, z)\| \,\mathrm{d}t \,.
$$

In the same vein as that to estimate I_2, and thanks to (4.52) and (4.51), we get

$$
J_1 \leq C\tau \ln \frac{1}{\tau} \|z\| \,.
$$

For J_2, by utilizing Lemmata 4.2 and 4.10, we have

$$
\begin{aligned}
J_2 &\leq \int_{(T-2\tau)\vee t_l}^{T} C\|z\| \,\mathrm{d}t + \int_{t_l}^{(T-2\tau)\vee t_l} \|\widehat{u}_{\pi(t)-1}(t_l, z) - u_h^*(t; t_l, z)\| \,\mathrm{d}t \\
&\leq C\tau\|z\| + C\tau\|z\| \int_{t_l}^{(T-2\tau)\vee t_l} \Big[\frac{\alpha}{T - \mu(t)} + \ln \frac{1}{\tau}\Big] \mathrm{d}t \\
&\leq C\tau \ln \frac{1}{\tau} \|z\| \,.
\end{aligned}
$$

Thanks to (4.52) in Lemma 4.9, we get

$$
J_3 \leq C\tau \Big[\frac{\alpha}{T - t_l} + \ln \frac{1}{\tau}\Big] \|z\| \,.
$$

A combination of these estimates then leads to

$$
0 \leq (\mathcal{P}_l \Pi_h z, z)_{\mathbb{L}^2} - \big(\mathcal{P}_h(t_l)\Pi_h z, z\big)_{\mathbb{L}^2} \leq C\|z\|^2 \tau \Big[\frac{\alpha}{T - t_l} + \ln \frac{1}{\tau}\Big], \tag{4.56}
$$

and estimates (4.55) and (4.56) together imply the assertion (4.54). □

b. The modified difference Riccati equation (4.57). In this part, we approximate the difference Riccati equation (4.20) to avoid computing $\mathcal{K}_n^{-1}$ for any $n = 0, 1, \ldots, N-1$, which is computationally expensive. In (4.20), by setting

$$\mathcal{V}_n = \mathcal{A}_0 \mathcal{P}_n \mathcal{A}_0 \qquad \forall n = 0, 1, \ldots, N-1 ,$$

and noticing the following identity

$$\begin{aligned}
\tau \mathcal{H}_n \mathcal{K}_n^{-1} \mathcal{H}_n &= \tau \mathcal{V}_{n+1} (\mathbb{1}_h + \tau \mathcal{V}_{n+1})^{-1} \mathcal{V}_{n+1} \\
&= \tau \mathcal{V}_{n+1} (\mathbb{1}_h + \tau \mathcal{V}_{n+1})^{-1} (\mathbb{1}_h + \tau \mathcal{V}_{n+1}) \mathcal{V}_{n+1} - \mathcal{V}_{n+1} (\mathbb{1}_h + \tau \mathcal{V}_{n+1})^{-1} \mathcal{V}_{n+1}^2 \tau^2 \\
&= \mathcal{V}_{n+1}^2 \tau - \mathcal{V}_{n+1} (\mathbb{1}_h + \tau \mathcal{V}_{n+1})^{-1} \mathcal{V}_{n+1}^2 \tau^2 ,
\end{aligned}$$

we can propose the following modification for (4.20) which results from dropping the $O(\tau^2)$-term

$$\begin{cases} \widetilde{\mathcal{P}}_n = \mathcal{A}_0 \widetilde{\mathcal{P}}_{n+1} \mathcal{A}_0 - \tau \mathcal{A}_0 \widetilde{\mathcal{P}}_{n+1} \mathcal{A}_0 \mathcal{A}_0 \widetilde{\mathcal{P}}_{n+1} \mathcal{A}_0 + \tau \mathbb{1}_h & n = 0, 1, \ldots, N-1 , \\ \widetilde{\mathcal{P}}_N = \alpha \mathbb{1}_h . \end{cases} \tag{4.57}$$

Note that to settle stability (as in Lemma 4.7) for this *modified difference Riccati equation* is not immediate, due to a missing related optimal control problem that established the estimate for (4.20). Instead, we use an induction argument here to verify that (4.57) is a rate-preserving perturbation of (4.20) which approximates (1.9); for this purpose, we introduce the difference

$$\delta \mathcal{P}_n \triangleq \mathcal{P}_n - \widetilde{\mathcal{P}}_n \qquad \forall n = 0, 1, \ldots, N .$$

By setting $\widetilde{\mathcal{V}}_n = \mathcal{A}_0 \widetilde{\mathcal{P}}_n \mathcal{A}_0$ and $\delta \mathcal{V}_n = \mathcal{V}_n - \widetilde{\mathcal{V}}_n$ for any $n = 0, 1, \ldots, N$, we find that

$$\begin{aligned}
\delta \mathcal{P}_n &= \delta \mathcal{V}_{n+1} - \mathcal{V}_{n+1} \delta \mathcal{V}_{n+1} \tau - \delta \mathcal{V}_{n+1} \widetilde{\mathcal{V}}_{n+1} \tau + \mathcal{V}_{n+1} (\mathbb{1}_h + \mathcal{V}_{n+1} \tau)^{-1} \mathcal{V}_{n+1}^2 \tau^2 \\
&= \delta \mathcal{V}_{n+1} - \mathcal{V}_{n+1} \delta \mathcal{V}_{n+1} \tau - \delta \mathcal{V}_{n+1} \mathcal{V}_{n+1} \tau + \delta \mathcal{V}_{n+1}^2 \tau \\
&\quad + \mathcal{V}_{n+1} (\mathbb{1}_h + \mathcal{V}_{n+1} \tau)^{-1} \mathcal{V}_{n+1}^2 \tau^2 .
\end{aligned}$$

Taking norms on both sides and noticing that $\|\mathcal{A}_0\|_{\mathcal{L}(\mathbb{L}^2|_{\mathbb{V}_h})} \leq 1$ and $\mathcal{P}_{n+1} \in \mathbb{S}_+(\mathbb{L}^2|_{\mathbb{V}_h})$, we arrive at

$$\begin{aligned}
\|\delta \mathcal{P}_n\|_{\mathcal{L}(\mathbb{L}^2|_{\mathbb{V}_h})} \leq{}& \|\delta \mathcal{P}_{n+1}\|_{\mathcal{L}(\mathbb{L}^2|_{\mathbb{V}_h})} + 2\|\mathcal{P}_{n+1}\|_{\mathcal{L}(\mathbb{L}^2|_{\mathbb{V}_h})} \|\delta \mathcal{P}_{n+1}\|_{\mathcal{L}(\mathbb{L}^2|_{\mathbb{V}_h})} \tau \\
&+ \|\delta \mathcal{P}_{n+1}\|_{\mathcal{L}(\mathbb{L}^2|_{\mathbb{V}_h})}^2 \tau + \|\mathcal{P}_{n+1}\|_{\mathcal{L}(\mathbb{L}^2|_{\mathbb{V}_h})}^3 \tau^2 .
\end{aligned} \tag{4.58}$$

Let

$$C_1 = 2 \max_{0 \leq n \leq N} \|\mathcal{P}_n\|_{\mathcal{L}(\mathbb{L}^2|_{\mathbb{V}_h})} \qquad C_2 = \max_{0 \leq n \leq N} \|\mathcal{P}_n\|_{\mathcal{L}(\mathbb{L}^2|_{\mathbb{V}_h})}^3 ,$$

which are both bounded by Lemma 4.7. Hence, (4.58) turns to

$$\|\delta \mathcal{P}_n\|_{\mathcal{L}(\mathbb{L}^2|_{\mathbb{V}_h})} \leq (1 + C_1 \tau) \|\delta \mathcal{P}_{n+1}\|_{\mathcal{L}(\mathbb{L}^2|_{\mathbb{V}_h})} + \|\delta \mathcal{P}_{n+1}\|_{\mathcal{L}(\mathbb{L}^2|_{\mathbb{V}_h})}^2 \tau + C_2 \tau^2 . \tag{4.59}$$

By setting

$$C_3 = 2(C_1 + C_2)\,, \tag{4.60}$$

we claim that: For τ small (will be specified later),

$$\|\delta\mathcal{P}_k\|_{\mathcal{L}(\mathbb{L}^2|_{\mathbb{V}_h})} \le e^{C_3(N-k)\tau}\tau \qquad \forall k = 0, 1, \ldots, N\,. \tag{4.61}$$

We show this claim by induction. For $k = N$, it is easy to see that

$$\|\delta\mathcal{P}_k\|_{\mathcal{L}(\mathbb{L}^2|_{\mathbb{V}_h})} = \|\delta\mathcal{P}_N\|_{\mathcal{L}(\mathbb{L}^2|_{\mathbb{V}_h})} = 0 \le e^{C_3(N-N)\tau}\tau\,.$$

Now, suppose that (4.61) holds for $k = N, N-1, \ldots, n+1$. Then (4.59) leads to

$$\begin{aligned}\|\delta\mathcal{P}_n\|_{\mathcal{L}(\mathbb{L}^2|_{\mathbb{V}_h})} &\le (1 + C_1\tau)e^{C_3(N-n-1)\tau}\tau + e^{2C_3(N-n-1)\tau}\tau^3 + C_2\tau^2\\ &\le e^{C_3(N-n)\tau}\tau e^{-C_3\tau}\Big[\big[1 + (C_1 + C_2)\tau\big] + e^{C_3T}\tau^2\Big]\\ &=: e^{C_3(N-n)\tau}\tau f(\tau)\,.\end{aligned} \tag{4.62}$$

To derive the claim (4.61), we only need to prove that, for some positive constant $\widetilde{\tau}$

$$f(\tau) \le 1 \qquad \forall \tau \in (0, \widetilde{\tau}]\,. \tag{4.63}$$

We see that $f(0) = 1$ and take the derivative

$$f'(\tau) \le e^{-C_3\tau}\Big[-C_3 + (C_1 + C_2) + 2\tau e^{C_3T}\Big] = e^{-C_3\tau}\Big[-(C_1 + C_2) + 2\tau e^{C_3T}\Big]\,.$$

We use (4.60) in the last equality. By the above estimate, if $\widetilde{\tau}$ satisfies $-(C_1 + C_2) + 2\tau e^{C_3T} = 0$, i.e.,

$$\widetilde{\tau} = \frac{C_1 + C_2}{2} e^{-2(C_1+C_2)T}\,,$$

then $f'(\tau) \le 0$ for any $\tau \in (0, \widetilde{\tau}]$, and subsequently (4.63) holds. Therefore, by (4.62), it follows that the claim (4.61) is settled for such value of τ.

Relying on (4.61), actually we have derived the following convergence rate for the modified difference Riccati equation (4.57).

Theorem 4.3 *Let $\tau \in (0, \widetilde{\tau}]$ be sufficiently small. Suppose that $\mathcal{P}_h(\cdot)$ resp. $\widetilde{\mathcal{P}}_\cdot$ are solutions to* (4.8) *resp.* (4.57). *Then*

$$\|\mathcal{P}_h(t_l)\Pi_h - \widetilde{\mathcal{P}}_l\Pi_h\|_{\mathcal{L}(\mathbb{L}^2)} \le C\tau\Big[\frac{\alpha}{T - t_l} + \ln\frac{1}{\tau}\Big] \qquad \forall l = 0, 1, \ldots, N-1\,. \tag{4.64}$$

Proof Based on (4.61), for small τ, it holds that

$$\|\mathcal{P}_l - \widetilde{\mathcal{P}}_l\|_{\mathcal{L}(\mathbb{L}^2|_{\mathbb{V}_h})} \le C\tau \qquad \forall l = 0, 1, \ldots, N-1\,,$$

which, together with Theorem 4.2 and the triangle inequality, yields the assertion (4.64). □

Remark 4.2 For $\mathcal{P}_\cdot$ from (4.19), on using the fact that weight coefficients in the related cost functional $\mathcal{G}_{h,\tau}(\cdot,\cdot;\cdot)$ are nonnegative, we can conclude that $\mathcal{P}_n \in \mathbb{S}_+(\mathbb{L}^2|_{\mathbb{V}_h})$ for $n = 0, 1, \ldots, N$. In contrast, for $\widetilde{\mathcal{P}}_\cdot$ from (4.57) *no* cost functional $\widetilde{\mathcal{G}}_{h,\tau}(\cdot,\cdot;\cdot)$ is available, and we may only deduce that $\widetilde{\mathcal{P}}_n \in \mathbb{S}(\mathbb{L}^2|_{\mathbb{V}_h})$ for $n = 0, 1, \ldots, N$. If one could obtain a positive lower bound for $\min_{0\le n\le N-1} \|\mathcal{P}_n\|_{\mathcal{L}(\mathbb{L}^2|_{\mathbb{V}_h})}$, then based on (4.61) the further property $\widetilde{\mathcal{P}}_n \in \mathbb{S}_+(\mathbb{L}^2|_{\mathbb{V}_h})$ for $n = 0, 1, \ldots, N$ could be derived for τ sufficiently small.

4.2.3 *The Second Consistent Difference Riccati Equation*

The error analysis in Sects. 4.2.1 and 4.2.2 for Problem $\mathbf{(LQ)}^t_{\rm aux}$ assume **(H)**—which is a non-checkable assumption in applications. Another limitation of Theorem 4.2 is that *no* convergence way be expected for the 'initial' time step $t_l = t_{N-1}$ unless $\alpha = 0$, since the estimate (4.54) at this time is of the form

$$\|\mathcal{P}_h(t_{N-1})\Pi_h - \mathcal{P}_{N-1}\Pi_h\|_{\mathcal{L}(\mathbb{L}^2)} \le C\Big[\alpha + \tau \ln \frac{1}{\tau}\Big],$$

such that the right-hand side does *not* tend to zero for $\tau \to 0$. However, the estimate (4.54) is sufficient to derive the convergence rate for the spatio-temporal discretization of Problem **(SLQ)**; see also [5, Theorem 4.5].

In this section, we present different spatio-temporal discretization schemes that are exempted from both deficiencies of the *difference Riccati equation* (4.20) above:

(a) General values for $\beta \in \mathbb{R}$ are allowed to scale noise. Hence, assumption **(H)** is *not* needed anymore.
(b) Theorems 4.5 and 4.6 will be shown, which even ensure convergence rates for iterates of the new *difference Riccati equations* (4.68) and (4.94) respectively.

a. The difference Riccati equation (4.68). In this part, we present another discretization for differential Riccati equation (4.8). Since we do *not* assume **(H)** any more, we rewrite Riccati equation (4.8) and system (4.11) by using Δ_h instead of $\mathcal{A}_h$ as follows

$$\begin{cases} \mathcal{P}_h'(t)+\Delta_h\mathcal{P}_h(t)+\mathcal{P}_h(t)\Delta_h+\beta^2\mathcal{P}_h(t)+\mathbb{1}_h-\mathcal{P}_h^2(t)=0 & t\in[0,T]\,,\\ \mathcal{P}_h(T)=\alpha\mathbb{1}_h\,, \end{cases} \tag{4.65}$$

and

$$\begin{cases} x_h'(s) = \Delta_h x_h(s) + \dfrac{\beta^2}{2}x_h(s) + \Pi_h u_h(s) & s\in(t,T]\,,\\ x_h(t) = \Pi_h z\,. \end{cases} \tag{4.66}$$

In this section, to simplify notations, we still apply some that are used in former parts.

Now we approximate system (4.66) and the cost functional (4.10) as follows: For any given $l = 0, 1, \ldots, N-1$, and $z \in \mathbb{L}^2$

$$\begin{cases} x_{n+1} = \big[1 + \frac{\beta^2\tau}{2}\big]A_0 x_n + \tau A_0 \Pi_h u_n \qquad n = l, l+1, \ldots, N-1, \\ x_l = \Pi_h z, \end{cases} \tag{4.67}$$

where $A_0 = \big(\mathbb{1}_h - \tau\Delta_h\big)^{-1}$, and the cost functional $\mathcal{G}_{h,\tau}(t_l, z; \cdot)$ is defined in (4.17). Also, we can propose a discretized version of the family of Problem $(\mathbf{LQ})^{t;h}_{\text{aux}}$ as that proposed in Sect. 4.1.2.

Problem $(\mathbf{LQ})^{t_l;h,\tau}_{\text{aux}}$. For any given $l = 0, 1, \ldots, N-1$ and $z \in \mathbb{L}^2$, search for $u^*_\cdot \in \mathbb{U}(t_l, T)$ such that

$$\mathcal{G}_{h,\tau}(t_l, z; u^*_\cdot) = \inf_{u_\cdot \in \mathbb{U}(t_l,T)} \mathcal{G}_{h,\tau}(t_l, z; u_\cdot) =: V_{h,\tau}(t_l, z).$$

Different from difference Riccati equation (4.20), we introduce the following one:

$$\begin{cases} \bar{\mathcal{P}}_n = \big[1 + \frac{\beta^2\tau}{2}\big]^2 A_0 \bar{\mathcal{P}}_{n+1} A_0 + \tau\mathbb{1}_h - \tau\bar{\mathcal{H}}_n\bar{\mathcal{K}}_n^{-1}\bar{\mathcal{H}}_n \qquad n = 0, 1, \ldots, N-1, \\ \bar{\mathcal{P}}_N = \alpha\mathbb{1}_h, \\ \bar{\mathcal{H}}_n = \big[1 + \frac{\beta^2\tau}{2}\big]A_0\bar{\mathcal{P}}_{n+1}A_0, \\ \bar{\mathcal{K}}_n = \mathbb{1}_h + \tau A_0\bar{\mathcal{P}}_{n+1}A_0, \quad \text{with} \quad A_0 = \big(\mathbb{1}_h - \tau\Delta_h\big)^{-1}. \end{cases} \tag{4.68}$$

As in Lemma 4.1, we can deduce that Problem $(\mathbf{LQ})^{t_l;h,\tau}_{\text{aux}}$ has a unique optimal control which enjoys a *discrete feedback form*

$$u^*_n = -\bar{\mathcal{K}}_n^{-1}\bar{\mathcal{H}}_n x^*_n \qquad n = l, l+1, \ldots, N-1, \tag{4.69}$$

and the *discrete value function* is given by

$$V_{h,\tau}(t_l, z) = \frac{1}{2}(\bar{\mathcal{P}}_l \Pi_h z, z)_{\mathbb{L}^2} \qquad \forall z \in \mathbb{L}^2. \tag{4.70}$$

By (4.69), or with the same argument as in (4.14), we deduce that the optimal control $u^*_\cdot$ to Problem $(\mathbf{LQ})^{t_l;h,\tau}_{\text{aux}}$ is $\mathbb{V}_h$-valued.

We may follow the procedure in Sect. 4.2.2 to derive the following results. We only provide proofs where the argumentation differs.

Lemma 4.11 *For any $l = 0, 1, \ldots, N-1$, $z \in \mathbb{L}^2$, $u_\cdot \in \mathbb{U}(t_l, T)$, and any $n = l, l+1, \ldots, N$, it holds that*

$$x_n(t_l, z, u.) = \big[1 + \frac{\beta^2\tau}{2}\big]^{n-l} A_0^{n-l}\Pi_h z + \tau \sum_{k=l}^{n-1} \big[1 + \frac{\beta^2\tau}{2}\big]^{n-k-1} A_0^{n-k}\Pi_h u_k \,, \tag{4.71}$$

or equivalently

$$x_n(t_l, z, u.) = A_0^{n-l}\Pi_h z + \tau \sum_{k=l}^{n-1} A_0^{n-k}\big[\frac{\beta^2}{2} x_k(t_l, z, u.) + \Pi_h u_k\big]. \tag{4.72}$$

Lemma 4.12 *Let $\bar{\mathcal{P}}. \equiv \{\bar{\mathcal{P}}_n\}_{n=0}^N \subset \mathbb{S}_+(\mathbb{L}^2|_{\mathbb{V}_h})$ be the solution to difference Riccati equation (4.68), and for any $l = 0, 1, \ldots, N-1$, $z \in \mathbb{L}^2$, let $\big(x_\cdot^*(t_l, z), u_\cdot^*(t_l, z)\big)$ be the optimal pair of Problem $(\mathbf{LQ})_{\mathrm{aux}}^{t_l;h,\tau}$. Then, there exists a constant C independent of h and τ such that*

$$\max_{0\le l\le N-1} \|\bar{\mathcal{P}}_l\|_{\mathcal{L}(\mathbb{L}^2|_{\mathbb{V}_h})} \le C\,, \tag{4.73}$$

$$\max_{0\le l\le N-1}\ \max_{l\le n\le N-1} \big[\|x_n^*(t_l, z)\| + \|u_n^*(t_l, z)\|\big] \le C\|z\|\,. \tag{4.74}$$

The following result states the discrete version of Pontryagin's maximum principle for Problem $(\mathbf{LQ})_{\mathrm{aux}}^{t_l;h,\tau}$; the proof is similar to that of Theorem 3.7.

Theorem 4.4 *For any $l = 0, 1, \ldots, N-1$, and $z \in \mathbb{L}^2$, the unique optimal pair $\big(x_\cdot^*(t_l, z), u_\cdot^*(t_l, z)\big) \in \mathbb{X}(t_l, T) \times \mathbb{U}(t_l, T)$ of Problem $(\mathbf{LQ})_{\mathrm{aux}}^{t_l;h,\tau}$ solves the following coupled equations for $n = l, l+1, \ldots, N-1$:*

$$\begin{cases} x_{n+1}^*(t_l, z) = \big[1 + \frac{\beta^2\tau}{2}\big]^{n+1-l} A_0^{n+1-l}\Pi_h z \\ \qquad\qquad + \tau \sum_{k=l}^{n} \big[1 + \frac{\beta^2\tau}{2}\big]^{n-k} A_0^{n+1-k}\Pi_h u_k^*(t_l, z)\,, \\ y_n = -\tau \sum_{k=n+1}^{N-1} \big[1 + \frac{\beta^2\tau}{2}\big]^{k-1-n} A_0^{k-n} x_k^*(t_l, z) \\ \qquad\qquad - \big[1 + \frac{\beta^2\tau}{2}\big]^{N-1-n} A_0^{N-n}\alpha x_N^*(t_l, z)\,, \\ x_l^*(t_l, z) = \Pi_h z\,, \end{cases}$$

together with the discrete optimality condition

$$u_n^*(t_l, z) - y_n = 0\,. \tag{4.75}$$

Remark 4.3 By Theorem 4.4, we can deduce that $y.$ solves the following difference equation

$$\begin{cases} y_{n+1} - y_n = -\tau\Delta_h y_n - \frac{\beta^2\tau}{2} y_{n+1} + \tau x_{n+1}^*(t_l, z)\,, \\ y_N = -\frac{\alpha - \tau}{1 + \frac{\beta^2}{2}\tau} x_N^*(t_l, z)\,. \end{cases} \tag{4.76}$$

Note that since the terminal condition $y_N \neq y_h(T)$, $y_\cdot$ is different from the semi-implicit Euler method for (4.28).

Lemma 4.13 *For any $l = 0, 1, \ldots, N-1$, $z \in \mathbb{L}^2$, suppose that $u^*_\cdot(t_l, z)$ is the optimal control of Problem* $(\mathbf{LQ})^{t_l;h,\tau}_{\mathtt{aux}}$. *Then there exists a constant C independent of h, τ such that*

$$\max_{0\le l\le N-1}\Big[\max_{l\le n\le N-1}\|\nabla u^*_n(t_l,z)\|^2 + \tau\sum_{n=l}^{N-1}\|\Delta_h u^*_n(t_l,z)\|^2\Big] \le C\|z\|^2_{\mathbb{H}^1_0}.$$

Proof Testing (4.76) by $\Delta_h y_n$, applying discrete Gronwall's inequality and then taking the summation, we deduce that

$$\begin{aligned}\max_{l\le n\le N-1}\|\nabla y_n\|^2 + \tau\sum_{n=l}^{N-1}\|\Delta_h y_n\|^2 &\le C\Big[\|\nabla y_N\|^2 + \tau\sum_{n=l+1}^{N}\|x^*_n(t_l,z)\|^2\Big]\\ &\le C\big[\|\nabla x^*_N(t_l,z)\|^2 + \max_{l\le n\le N}\|x^*_n(t_l,z)\|^2\big].\end{aligned}$$

Similarly if we test (4.67) with $\Delta_h x^*_{n+1}(t_l,z)$, then by (4.74) in Lemma 4.12

$$\begin{aligned}&\max_{l\le n\le N}\|\nabla x^*_n(t_l,z)\|^2 + \tau\sum_{n=l}^{N-1}\|\Delta_h x^*_{n+1}(t_l,z)\|^2\\ &\le C\Big[\|\nabla x^*_l(t_l,z)\|^2 + \tau\sum_{n=l}^{N-1}\|u^*_n(t_l,z)\|^2\Big]\\ &\le C\|z\|^2_{\mathbb{H}^1_0}.\end{aligned}$$

By combining these two estimates with the optimality condition (4.75), we conclude that

$$\max_{l\le n\le N-1}\|\nabla u^*_n(t_l,z)\|^2 + \tau\sum_{n=l}^{N-1}\|\Delta_h u^*_n(t_l,z)\|^2 \le C\|z\|^2_{\mathbb{H}^1_0},$$

where C is independent of l. That settles the assertion. □

Lemma 4.14 *For any $l = 0, 1, \ldots, N-1$ and $z \in \mathbb{H}^1_0$, let $\big(x^*_\cdot(t_l,z), u^*_\cdot(t_l,z)\big)$ be the optimal pair of Problem* $(\mathbf{LQ})^{t_l;h,\tau}_{\mathtt{aux}}$. *Suppose that*

$$\bar u_h(t;t_l,z) = u^*_n(t_l,z) \qquad \forall t\in[t_n,t_{n+1}),\ n = l, l+1, \ldots, N-1, \tag{4.77}$$

and $\bar x_h(\cdot;t_l,z) \triangleq x_h\big(\cdot;t_l,z,\bar u_h(\cdot;t_l,z)\big)$. There exists a constant C independent of h, τ such that

$$\max_{0\le l\le N-1}\Big[\sup_{t\in[t_l,T]}\|\bar{x}_h(t;t_l,z)\|^2_{\mathbb{H}^1_0}+\int_{t_l}^T\|\Delta_h\bar{x}_h(t;t_l,z)\|^2\,dt\Big]\le C\|z\|_{\mathbb{H}^1_0},\tag{4.78}$$

$$\sum_{n=l}^{N-1}\int_{t_n}^{t_{n+1}}\big[\|\bar{x}_h(t;t_l,z)-\bar{x}_h(t_n;t_l,z)\|+\|\bar{x}_h(t;t_l,z)-\bar{x}_h(t_{n+1};t_l,z)\|\big]\,dt$$
$$\le C\tau\|z\|_{\mathbb{H}^1_0},\tag{4.79}$$

$$\|\bar{x}_h(t_n;t_l,z)-x^*_n(t_l,z)\|\le C\frac{\tau}{\sqrt{t_n-t_l}}\|z\|_{\mathbb{H}^1_0}\quad n=l+1,\cdots,N.\tag{4.80}$$

Proof The proof is similar to that of Lemma 4.8. For simplicity, we write $\bar{x}_h(\cdot)$ for $\bar{x}_h(\cdot;t_l,z)$ throughout the proof.

(1) Verification of (4.78). A standard stability estimate for the solution to equation (4.66), and Lemma 4.12 lead to

$$\begin{aligned}&\sup_{t\in[t_l,T]}\|\bar{x}_h(t;t_l,z)\|^2_{\mathbb{H}^1_0}+\int_{t_l}^T\|\Delta_h\bar{x}_h(t;t_l,z)\|^2\,\mathrm{d}t\\&\le C\Big[\|\Pi_h z\|^2_{\mathbb{H}^1_0}+\int_{t_l}^T\|\bar{u}_h(t;t_l,z)\|^2\,\mathrm{d}t\Big]\\&\le C\|z\|^2_{\mathbb{H}^1_0},\end{aligned}$$

where C is independent of l. That settles the assertion.

(2) Verification of (4.79). We only prove

$$I:=\sum_{n=l}^{N-1}\int_{t_n}^{t_{n+1}}\|\bar{x}_h(t)-\bar{x}_h(t_n)\|\,\mathrm{d}t\le C\tau\|z\|_{\mathbb{H}^1_0},$$

and the remaining part can be obtained similarly. By (4.66), we find that

$$\begin{aligned}I&\le\sum_{n=l}^{N-1}\int_{t_n}^{t_{n+1}}\big\|\big[E_h(t-t_l)-E_h(t_n-t_l)\big]\Pi_h z\big\|\,\mathrm{d}t\\&\quad+\sum_{n=l}^{N-1}\int_{t_n}^{t_{n+1}}\Big\|\int_{t_l}^{t_n}\big[E_h(t-\theta)-E_h(t_n-\theta)\big]\big[\frac{\beta^2}{2}\bar{x}_h(\theta)+\bar{u}_h(\theta,t_l,z)\big]\,\mathrm{d}\theta\Big\|\,\mathrm{d}t\\&\quad+\sum_{n=l}^{N-1}\int_{t_n}^{t_{n+1}}\Big\|\int_{t_n}^{t}E_h(t-\theta)\big[\frac{\beta^2}{2}\bar{x}_h(\theta)+u^*_n(t_l,z)\big]\,\mathrm{d}\theta\Big\|\,\mathrm{d}t\\&=:I_1+I_2+I_3.\end{aligned}\tag{4.81}$$

Now we estimate summands on the right-hand side of (4.81). For I_1, by virtue of $\nu(\cdot)$ defined in (2.8), Remark 2.1, and (2.2c) with $\gamma=0$ as well as (2.2a) with $\gamma=1/2$, (2.2b) with $\gamma=1$ in Lemma 2.1, it follows that

$$I_1 \leq \int_{t_l}^{t_{l+1}} C\|z\|\,\mathrm{d}t + \int_{t_{l+1}}^{T} C\frac{t-\nu(t)}{\sqrt{t-\tau-t_l}}\|z\|_{\mathbb{H}_0^1}\,\mathrm{d}t \leq C\tau\|z\|_{\mathbb{H}_0^1}\,. \tag{4.82}$$

For I_2, by Lemma 2.1, Remark 2.1, assertion (4.78) and Lemma 4.13, it follows that

$$I_2 \leq C\sum_{n=l}^{N-1}\int_{t_n}^{t_{n+1}}\int_{t_l}^{t_n}\tau\big[\|\Delta_h\bar{x}_h(\theta)\| + \|\Delta_h\bar{u}_h(\theta;t_l,z)\|\big]\,\mathrm{d}\theta\,\mathrm{d}t \leq C\tau\|z\|_{\mathbb{H}_0^1}\,. \tag{4.83}$$

For I_3, by (4.74) and (4.78), it is easy to see that

$$I_3 \leq C\tau\Big[\sup_{t\in[t_l,T]}\|\bar{x}_h(t)\| + \max_{l\leq n\leq N-1}\|u_n^*(t_l,z)\|\Big] \leq C\tau\|z\|_{\mathbb{H}_0^1}\,. \tag{4.84}$$

Combining with (4.81)–(4.84), we prove the second assertion.

(3) Verification of (4.80). By (4.66) and (4.67), we find that

$$\begin{aligned}
\|\bar{x}_h(t_{n+1}) - x_{n+1}^*(t_l,z)\| \leq\; & \big\|E_h(t_{n+1}-t_l)\Pi_h z - A_0^{n+1-l}\Pi_h z\big\| \\
& + \Big\|\sum_{j=l}^{n}\int_{t_j}^{t_{j+1}}\big[E_h(t_{n+1}-t) - A_0^{n+1-j}\Pi_h\big]u_j^*(t_l,z)\,\mathrm{d}t\Big\| \\
& + \frac{\beta^2}{2}\Big\|\sum_{j=l}^{n}\int_{t_j}^{t_{j+1}}\big[E_h(t_{n+1}-t) - A_0^{n+1-j}\Pi_h\big]\bar{x}_h(t)\,\mathrm{d}t\Big\| \\
& + \frac{\beta^2}{2}\Big\|\sum_{j=l}^{n}\int_{t_j}^{t_{j+1}}A_0^{n+1-j}\Pi_h\big[\bar{x}_h(t) - \bar{x}_h(t_j)\big]\,\mathrm{d}t\Big\| \\
& + \frac{\beta^2}{2}\Big\|\sum_{j=l}^{n}\int_{t_j}^{t_{j+1}}A_0^{n+1-j}\Pi_h\big[\bar{x}_h(t_j) - x_j^*(t_l,z)\big]\,\mathrm{d}t\Big\| \\
=:\; & \sum_{i=1}^{5}J_i\,.
\end{aligned}$$

Now, we estimate these five terms one by one. By (2.10b) with $\gamma = 2$ in Lemma 2.4 and (2.2c) in Lemma 2.1 with $\gamma = 1$, following the vein to estimate J_1 in the proof of Lemma 4.8, we conclude that

$$J_1 \leq C\frac{\tau}{\sqrt{t_{n+1}-t_l}}\|z\|_{\mathbb{H}_0^1}\,.$$

Also, by (2.10b) with $\gamma = 2$ in Lemma 2.4 and Lemma 4.13,

$$J_2 \leq \int_{t_l}^{t_{n+1}}\|G_\tau(t_{n+1}-t)\bar{u}_h(t;t_l,z)\|\,\mathrm{d}t \leq C\tau\|z\|_{\mathbb{H}_0^1}\,.$$

By the same trick, applying (4.78), we can deduce that

$$J_3 \le C\tau \|z\|_{\mathbb{H}_0^1} .$$

By (4.79), it is evident that

$$J_4 \le C \sum_{j=l}^{n} \int_{t_j}^{t_{j+1}} \|\bar{x}_h(t) - \bar{x}_h(t_j)\| \, \mathrm{d}t \le C\tau \|z\|_{\mathbb{H}_0^1} .$$

For J_5, we can easily see that

$$J_5 \le C\tau \sum_{j=l}^{n} \|\bar{x}_h(t_j) - x_j^*(t_l, z)\| = C\tau \sum_{j=l+1}^{n} \|\bar{x}_h(t_j) - x_j^*(t_l, z)\| .$$

By setting $e_n = \|\bar{x}_h(t_n) - x_n^*(t_l, z)\|$ and combining with above estimates, we have

$$e_{n+1} \le C \frac{\tau}{\sqrt{t_{n+1} - t_l}} \|z\|_{\mathbb{H}_0^1} + C\tau \sum_{j=l+1}^{n} e_j ,$$

which, together with discrete Gronwall's inequality, yields

$$e_{n+1} \le C \frac{\tau}{\sqrt{t_{n+1} - t_l}} \|z\|_{\mathbb{H}_0^1} + C\tau \sum_{j=l+1}^{n} \frac{\tau}{\sqrt{t_j - t_l}} \|z\|_{\mathbb{H}_0^1} \le C \frac{\tau}{\sqrt{t_{n+1} - t_l}} \|z\|_{\mathbb{H}_0^1} .$$

That completes the proof. □

The proof of the following regularity result on $u_h^*(\cdot; t, z)$ can be derived in the same vein as that of Lemma 4.13, and we omit it.

Lemma 4.15 *For any $t \in [0, T)$, $z \in \mathbb{L}^2$, suppose that $u_h^*(\cdot; t, z)$ is the optimal control of Problem* $\textbf{(LQ)}_{\mathrm{aux}}^{t;h}$. *Then there exists a constant C independent of h such that*

$$\sup_{s \in [t,T]} \|\nabla u_h^*(s; t, z)\|^2 + \int_t^T \|\Delta_h u_h^*(s; t, z)\|^2 \, ds \le C \|z\|_{\mathbb{H}_0^1}^2 .$$

The following result is similar to Lemma 4.9 and can be proved similarly to Lemma 4.14.

Lemma 4.16 *For any $l = 0, 1, \ldots, N-1$ and $z \in \mathbb{L}^2$, let $\big(x_h^*(\cdot; t_l, z), u_h^*(\cdot; t_l, z)\big)$ be the optimal pair of Problem* $\textbf{(LQ)}_{\mathrm{aux}}^{t_l;h}$. *Suppose that*

$$\widehat{u}_n(t_l, z) = \frac{1}{\tau} \int_{t_n}^{t_{n+1}} u_h^*(t; t_l, z) \, dt \qquad \forall n = l, l+1, \ldots, N-1 , \tag{4.85}$$

and $\widehat{x}_\cdot(t_l, z) \triangleq x_\cdot\big(t_l, z, \widehat{u}_\cdot(t_l, z)\big)$. *Then*

$$\max_{0\le l\le N-1}\Big[\max_{l\le n\le N}\|\nabla\widehat{x}_n(t_l,z)\|^2+\tau\sum_{n=l}^{N-1}\|\Delta_h\widehat{x}_n(t_l,z)\|^2\Big]\le C\|z\|^2_{\mathbb{H}^1_0}\,, \tag{4.86}$$

$$\int_{t_l}^{T}\Big[\big\|x^*_h(t;t_l,z)-x^*_h\big(\nu(t);t_l,z\big)\big\|+\big\|x^*_h(t;t_l,z)-x^*_h\big(\mu(t);t_l,z\big)\big\|\Big]dt \le C\tau\|z\|_{\mathbb{H}^1_0}\,, \tag{4.87}$$

$$\|\widehat{x}_n(t_l,z)-x^*_h(t_n;t_l,z)\|\le C\frac{\tau}{\sqrt{t_n-t_l}}\|z\|_{\mathbb{H}^1_0}\quad n=l+1,\cdots,N\,. \tag{4.88}$$

Proof (1) Verification of (4.86). Derived by Lemma 4.15 and the same vein as in the proof of Lemma 4.13.

(2) Verification of (4.87) and (4.88). As done for (4.79) and (4.80). □

The following result can be proved in the same way as that in the proof of Lemma 4.10.

Lemma 4.17 *For a given uniform partition* I_τ *of size* $\tau\in(0,\tau_0]\subset(0,1)$, *for any* $l=0,1,\ldots,N-1$ *and* $z\in\mathbb{L}^2$, *suppose that* $u^*_h(\cdot;t_l,z)$ *is the optimal control of Problem* $\mathbf{(LQ)}^{t_l;h}_{\rm aux}$. *Then, for* $s,s_0\in[t_n,t_{n+1})$, $n=l,l+1,\ldots,N-1$, *there exists a constant* C *independent of* τ *such that*

$$\|u^*_h(s;t_l,z)-u^*_h(s_0;t_l,z)\|\le C\tau\frac{1}{\sqrt{T-s\vee s_0}}\|z\|_{\mathbb{H}^1_0}\,. \tag{4.89}$$

Now we are in the position to bound the error of $\bar{\mathcal{P}}_l-\mathcal{P}_h(t_l)$.

Theorem 4.5 *Suppose that* $\mathcal{P}_h(\cdot)$ *and* $\bar{\mathcal{P}}_\cdot$ *are solutions to Riccati equations* (4.65) *and* (4.68) *respectively. Then there exists a constant* C *independent of* h *and* τ *such that*

$$\|\mathcal{P}_h(t_l)\Pi_h-\bar{\mathcal{P}}_l\Pi_h\|_{\mathcal{L}(\mathbb{H}^1_0;\mathbb{L}^2)}\le C\tau\Big[\frac{\alpha}{\sqrt{T-t_l}}+1\Big]\quad l=0,1,\ldots,N-1\,. \tag{4.90}$$

Proof The proof is long and we divide it into two steps.
(1) In this step, we show that there exists a constant C such that for any $z\in\mathbb{H}^1_0$, $l=0,1,\ldots,N-1$,

$$\big|\big(\bar{\mathcal{P}}_l\Pi_hz,z\big)_{\mathbb{L}^2}-\big(\mathcal{P}_h(t_l)\Pi_hz,z\big)_{\mathbb{L}^2}\big|\le C\tau\Big[\frac{\alpha}{\sqrt{T-t_l}}+1\Big]\|z\|\|z\|_{\mathbb{H}^1_0}\,. \tag{4.91}$$

To do that, we proceed as in the proof of Theorem 4.2. For given $l=0,1,\ldots,N-1$, and $z\in\mathbb{H}^1_0$, we divide the proof into the following two cases.

Case (i). $\big(\bar{\mathcal{P}}_l\Pi_hz,z\big)_{\mathbb{L}^2}\le\big(\mathcal{P}_h(t_l)\Pi_hz,z\big)_{\mathbb{L}^2}$.

By relying on Lemmata 4.12 and 4.14, we can conclude that

$$0 \leq \big(\mathcal{P}_h(t_l)\Pi_h z, z\big)_{\mathbb{L}^2} - \big(\bar{\mathcal{P}}_l \Pi_h z, z\big)_{\mathbb{L}^2} \leq C\tau\Big[\frac{\alpha}{\sqrt{T-t_l}} + 1\Big]\|z\|\|z\|_{\mathbb{H}_0^1}.$$

Case (ii). $\big(\bar{\mathcal{P}}_l \Pi_h z, z\big)_{\mathbb{L}^2} > \big(\mathcal{P}_h(t_l)\Pi_h z, z\big)_{\mathbb{L}^2}$.

In the same vein as in Case (ii) of Theorem 4.2, by Lemmata 4.16 and 4.17, we can deduce that

$$\begin{aligned}
0 &\leq \big(\bar{\mathcal{P}}_l \Pi_h z, z\big)_{\mathbb{L}^2} - \big(\mathcal{P}_h(t_l)\Pi_h z, z\big)_{\mathbb{L}^2} \\
&\leq C\|z\|\Big\{\sum_{n=l}^{N-1}\Big[\int_{t_n}^{t_{n+1}} \|\widehat{x}_n(t_l, z) - x_h^*(t_n; t_l, z)\| + \|x_h^*(t_n; t_l, z) - x_h^*(t; t_l, z)\| \\
&\quad + \Big\|\frac{1}{\tau}\int_{t_n}^{t_{n+1}} u_h^*(s; t_l, z)\,\mathrm{d}s - u_h^*(t; t_l, z)\Big\|\Big]\,\mathrm{d}t + \alpha\|\widehat{x}_N(t_l, z) - x_h^*(T; t_l, z)\|\Big\} \\
&=: C\|z\|\sum_{i=1}^{4} J_i.
\end{aligned}$$

For J_1, (4.88) in Lemma 4.16 and the triangle inequality yield

$$J_1 \leq C\tau\|z\| + C\sum_{n=l+2}^{N-1}\int_{t_n}^{t_{n+1}} \frac{1}{\sqrt{t-\tau-t_l}}\|z\|_{\mathbb{H}_0^1}\,\mathrm{d}t \leq C\tau\|z\|_{\mathbb{H}_0^1}.$$

(4.87) and (4.88) in Lemma 4.16 lead to

$$J_2 \leq C\tau\|z\|_{\mathbb{H}_0^1}, \qquad J_4 \leq C\tau\frac{\alpha}{\sqrt{T-t_l}}\|z\|_{\mathbb{H}_0^1}.$$

Lemma 4.17 implies that

$$J_3 \leq \frac{2}{\tau}\sum_{n=l}^{N-1}\int_{t_n}^{t_{n+1}}\int_{t_n}^{s} \|u_h^*(s; t_l, z) - u_h^*(t; t_l, z)\|\,\mathrm{d}t\,\mathrm{d}s \leq C\tau\|z\|_{\mathbb{H}_0^1}.$$

A combination of these estimates leads to

$$0 \leq \big(\bar{\mathcal{P}}_l \Pi_h z, z\big)_{\mathbb{L}^2} - \big(\mathcal{P}_h(t_l)\Pi_h z, z\big)_{\mathbb{L}^2} \leq C\tau\Big[\frac{\alpha}{\sqrt{T-t_l}} + 1\Big]\|z\|\|z\|_{\mathbb{H}_0^1}.$$

Combining these two cases, we get the desired assertion.

(2) In this step, we adopt a spectral decomposition technique to derive sharp estimates. Based on the eigenvalues $\{\lambda_{h,i}\}_{i=1}^{\dim(\mathbb{V}_h)}$ resp. ($\mathbb{L}^2$-orthonormal) eigenfunctions $\{\varphi_{h,i}\}_{i=1}^{\dim(\mathbb{V}_h)}$, of the symmetric operator $-\Delta_h$, for any $i = 1, 2, \ldots, \dim(\mathbb{V}_h)$, we define a family of LQ problems.

Problem $(\mathbf{LQ})_{\mathrm{aux}}^{t;h,i}$. For any given $t \in [0, T)$ and $x \in \mathbb{R}$, search for $u_{h,i}^*(\cdot) \in L^2(t, T)$ such that

$$\mathcal{G}_h^i\big(t, x; u_{h,i}^*(\cdot)\big) = \inf_{u_h(\cdot)\in L^2(t,T)} \mathcal{G}_h^i\big(t, x; u_h(\cdot)\big) =: V_h^i(t, x)\,,$$

where the cost functional is

$$\mathcal{G}_h^i\big(t, x; u_h(\cdot)\big) = \frac{1}{2}\int_t^T \big[|x_{h,i}(s)|^2 + |u_h(s)|^2\big]\,\mathrm{d}s + \frac{\alpha}{2}|x_{h,i}(T)|^2\,,$$

and the state variable $x_{h,i}(\cdot)$ satisfies

$$\begin{cases} x_{h,i}'(s) = -\lambda_{h,i}x_{h,i}(s) + \dfrac{\beta^2}{2}x_{h,i}(s) + u_h(s) & s \in (t, T]\,,\\ x_{h,i}(t) = x \in \mathbb{R}\,. \end{cases}$$

By LQ theory, Problem $(\mathbf{LQ})_{\mathrm{aux}}^{t;h,i}$ has a unique optimal pair $\big(x_{h,i}^*(\cdot), u_{h,i}^*(\cdot)\big)$, satisfying the following state feedback representation:

$$u_{h,i}^*(s) = -p_{h,i}(s)x_{h,i}^*(s) \qquad s \in [t, T]\,,$$

where $p_{h,i}(\cdot)$ solves the following scalar Riccati equation

$$\begin{cases} p_{h,i}'(t) - 2\lambda_{h,i}p_{h,i}(t) + \beta^2 p_{h,i}(t) + 1 - p_{h,i}^2(t) = 0 & t \in [0, T]\,,\\ p_{h,i}(T) = \alpha\,. \end{cases}$$

By the uniqueness of the optimal controls to Problems $(\mathbf{LQ})_{\mathrm{aux}}^{t;h}$ and $(\mathbf{LQ})_{\mathrm{aux}}^{t;h,i}$, we conclude for any $z \in \mathbb{L}^2$ with $\Pi_h z = \sum_{i=1}^{\dim(\mathbb{V}_h)} (\Pi_h z, \varphi_{h,i})_{\mathbb{L}^2}\varphi_{h,i} =: \sum_{i=1}^{\dim(\mathbb{V}_h)} a_i\varphi_{h,i}$ that

$$x_h^*(\cdot; t, z) = \sum_{i=1}^{\dim(\mathbb{V}_h)} x_{h,i}^*\big(\cdot; t, a_i\big)\varphi_{h,i}\,, \qquad u_h^*(\cdot; t, z) = \sum_{i=1}^{\dim(\mathbb{V}_h)} u_{h,i}^*\big(\cdot; t, a_i\big)\varphi_{h,i}\,,$$

and

$$\mathcal{P}_h(\cdot)\varphi_{h,i} = p_{h,i}(\cdot)\varphi_{h,i} \qquad i = 1, 2, \ldots, \dim(\mathbb{V}_h)\,.$$

Similarly, we can define Problem $(\mathbf{LQ})_{\mathrm{aux}}^{t_l;h,\tau,i}$, derive $p_{\cdot,i}$, and for the optimal pair $(x_{\cdot,i}^*, u_{\cdot,i}^*)$

$$x_\cdot^*(t_l, z) = \sum_{i=1}^{\dim(\mathbb{V}_h)} x_{\cdot,i}^*\big(t_l, a_i\big)\varphi_{h,i}\,, \qquad u_\cdot^*(t_l, z) = \sum_{i=1}^{\dim(\mathbb{V}_h)} u_{\cdot,i}^*\big(t_l, a_i\big)\varphi_{h,i}\,,$$

and

$$\bar{\mathcal{P}}_{\cdot}\varphi_{h,i} = p_{\cdot,i}\varphi_{h,i} \qquad i = 1, 2, \ldots, \dim(\mathbb{V}_h)\,.$$

as well as

$$\max_{0\le l\le N-1}\ \max_{l\le n\le N} |p_{n,i}| \le C\,. \tag{4.92}$$

By taking $z = \varphi_{h,i}$ in (4.91), we have

$$\begin{aligned}
|p_{l,i} - p_{h,i}(t_l)| &= \big|(p_{l,i}\varphi_{h,i}, \varphi_{h,i})_{\mathbb{L}^2} - \big(p_{h,i}(t_l)\varphi_{h,i}, \varphi_{h,i}\big)_{\mathbb{L}^2}\big| \\
&= \big|\big(\bar{\mathcal{P}}_l\Pi_h\varphi_{h,i}, \varphi_{h,i}\big)_{\mathbb{L}^2} - \big(\mathcal{P}_h(t_l)\Pi_h\varphi_{h,i}, \varphi_{h,i}\big)_{\mathbb{L}^2}\big| \\
&\le C\lambda_{h,i}^{1/2}\tau\Big[\frac{\alpha}{\sqrt{T-t_l}} + 1\Big].
\end{aligned} \tag{4.93}$$

Thus for any $z \in \mathbb{H}_0^1$, we can deduce that

$$\begin{aligned}
\big\|\big[\bar{\mathcal{P}}_l\Pi_h - \mathcal{P}_h(t_l)\Pi_h\big]z\big\|^2 &= \Big\|\sum_{i=1}^{\dim(\mathbb{V}_h)} a_i\big[\bar{\mathcal{P}}_l\Pi_h - \mathcal{P}_h(t_l)\Pi_h\big]\varphi_{h,i}\Big\|^2 \\
&\le C\sum_{i=1}^{\dim(\mathbb{V}_h)} a_i^2\lambda_{h,i}\tau^2\Big[\frac{\alpha}{\sqrt{T-t_l}} + 1\Big]^2 \\
&\le C\|z\|_{\mathbb{H}_0^1}^2\tau^2\Big[\frac{\alpha}{\sqrt{T-t_l}} + 1\Big]^2,
\end{aligned}$$

which settles the assertion (4.90). □

b. The modified difference Riccati equation (4.94). In the below, we approximate the difference Riccati equation (4.68) to avoid computing $\bar{\mathcal{K}}_n^{-1}$ for any $n = 0, 1, \ldots, N-1$. Similar to (4.57), we perturb (4.68) by the following equation

$$\begin{cases}
\widehat{\mathcal{P}}_n = \big[1 + \frac{\beta^2\tau}{2}\big]^2\Big[A_0\widehat{\mathcal{P}}_{n+1}A_0 - \tau A_0\widehat{\mathcal{P}}_{n+1}A_0A_0\widehat{\mathcal{P}}_{n+1}A_0\Big] + \tau\mathbb{1}_h \\
\qquad\qquad n = 0, 1, \ldots, N-1\,, \\
\widehat{\mathcal{P}}_N = \alpha\mathbb{1}_h\,.
\end{cases} \tag{4.94}$$

Note that (4.94) is of the form (4.57)—apart from the new scaling factor of the leading term on the right-hand side. To measure the difference, we introduce

$$\delta\mathcal{P}_n \triangleq \bar{\mathcal{P}}_n - \widehat{\mathcal{P}}_n \qquad \forall n = 0, 1, \ldots, N\,.$$

By setting $\bar{\mathcal{V}}_n = A_0\bar{\mathcal{P}}_nA_0\,,\ \widehat{\mathcal{V}}_n = A_0\widehat{\mathcal{P}}_nA_0\,,\ \delta\mathcal{V}_n = \mathcal{V}_n - \widehat{\mathcal{V}}_n$ for any $n = 0, 1, \ldots, N$, we find that

$$\delta\mathcal{P}_n = \Big[1+\frac{\beta^2\tau}{2}\Big]^2\Big[\delta\mathcal{V}_{n+1} - \bar{\mathcal{V}}_{n+1}\delta\mathcal{V}_{n+1}\tau - \delta\mathcal{V}_{n+1}\bar{\mathcal{V}}_{n+1}\tau + \delta\mathcal{V}^2_{n+1}\tau + \bar{\mathcal{V}}_{n+1}(\mathbb{1}_h + \bar{\mathcal{V}}_{n+1}\tau)^{-1}\bar{\mathcal{V}}^2_{n+1}\tau^2\Big]. \tag{4.95}$$

With the same procedure as that to derive (4.61), we can arrive at

$$\max_{0\le n\le N} \|\delta\mathcal{P}_n\|_{\mathcal{L}(\mathbb{L}^2|_{\mathbb{V}_h})} \le C\tau\,. \tag{4.96}$$

By virtue of $\{p_{\cdot,i}\}_{i=1}^{\dim(\mathbb{V}_h)}$, for any $\Pi_h z = \sum_{i=1}^{\dim(\mathbb{V}_h)} a_i\varphi_{h,i}$

$$\begin{aligned}\|\bar{\mathcal{P}}_n\Pi_h z\|^2 &\le \|\bar{\mathcal{P}}_n\Pi_h z\|^2_{\mathbb{H}^1_0} = \Big\|\sum_{i=1}^{\dim(\mathbb{V}_h)} a_i p_{n,i}\varphi_{h,i}\Big\|^2_{\mathbb{H}^1_0}\\ &= \sum_{i=1}^{\dim(\mathbb{V}_h)} a_i^2\lambda_{h,i}|p_{n,i}|^2 \le C\|\nabla\Pi_h z\|^2 \le C\|z\|^2_{\mathbb{H}^1_0}\,,\end{aligned}$$

which leads to $\|\bar{\mathcal{P}}_n\|_{\mathcal{L}(\mathbb{H}^1_0|_{\mathbb{V}_h};\mathbb{L}^2|_{\mathbb{V}_h})} \le \|\bar{\mathcal{P}}_n\|_{\mathcal{L}(\mathbb{H}^1_0|_{\mathbb{V}_h})}$ and $\max_{0\le n\le N}\|\bar{\mathcal{P}}_n\|_{\mathcal{L}(\mathbb{H}^1_0|_{\mathbb{V}_h})} \le C$. Taking $\mathcal{L}(\mathbb{H}^1_0|_{\mathbb{V}_h}, \mathbb{L}^2|_{\mathbb{V}_h})$-norm on both sides of (4.95) and applying the fact that $\|A_0\|_{\mathcal{L}(\mathbb{L}^2|_{\mathbb{V}_h})} \le 1$, $\|A_0\|_{\mathcal{L}(\mathbb{H}^1_0|_{\mathbb{V}_h})} \le 1$, we can deduce that

$$\begin{aligned}&\|\delta\mathcal{P}_n\|_{\mathcal{L}(\mathbb{H}^1_0|_{\mathbb{V}_h};\mathbb{L}^2|_{\mathbb{V}_h})}\\ &\le \Big[1+\frac{\beta^2\tau}{2}\Big]^2\Big[\|\delta\mathcal{P}_{n+1}\|_{\mathcal{L}(\mathbb{H}^1_0|_{\mathbb{V}_h};\mathbb{L}^2|_{\mathbb{V}_h})} + \|\bar{\mathcal{P}}_{n+1}\|_{\mathcal{L}(\mathbb{L}^2|_{\mathbb{V}_h})}\|\delta\mathcal{P}_{n+1}\|_{\mathcal{L}(\mathbb{H}^1_0|_{\mathbb{V}_h};\mathbb{L}^2|_{\mathbb{V}_h})}\tau\\ &\quad+\|\delta\mathcal{P}_{n+1}\|_{\mathcal{L}(\mathbb{H}^1_0|_{\mathbb{V}_h};\mathbb{L}^2|_{\mathbb{V}_h})}\|\bar{\mathcal{P}}_{n+1}\|_{\mathcal{L}(\mathbb{H}^1_0|_{\mathbb{V}_h})}\tau\\ &\quad+\|\delta\mathcal{P}_{n+1}\|_{\mathcal{L}(\mathbb{L}^2|_{\mathbb{V}_h})}\|\delta\mathcal{P}_{n+1}\|_{\mathcal{L}(\mathbb{H}^1_0|_{\mathbb{V}_h};\mathbb{L}^2|_{\mathbb{V}_h})}\tau + \|\bar{\mathcal{P}}_{n+1}\|^3_{\mathcal{L}(\mathbb{L}^2|_{\mathbb{V}_h})}\tau^2\Big]\\ &\le (1+C\tau)\|\delta\mathcal{P}_{n+1}\|_{\mathcal{L}(\mathbb{H}^1_0|_{\mathbb{V}_h};\mathbb{L}^2|_{\mathbb{V}_h})} + C\tau^2\,,\end{aligned}$$

which, together with discrete Gronwall's inequality, yields

$$\max_{0\le n\le N} \|\delta\mathcal{P}_n\|_{\mathcal{L}(\mathbb{H}^1_0|_{\mathbb{V}_h};\mathbb{L}^2)} \le C\tau\,. \tag{4.97}$$

Relying on (4.97) and Theorem 4.5, we actually have derived the same convergence rate for the *modified difference Riccati equation* (4.94) as in Theorem 4.5 for the difference Riccati equation (4.68). Since the proof is similar to that of Theorem 4.3, we omit it.

Theorem 4.6 *Let* $\tau \in (0, \widetilde{\tau}]$ *be sufficiently small. Suppose that* $\mathcal{P}_h(\cdot)$ *resp.* $\widehat{\mathcal{P}}_\cdot$ *are solutions to* (4.65) *resp.* (4.94). *Then there exists a constant C independent of h and* τ *such that*

$$\|\mathcal{P}_h(t_l)\Pi_h - \widehat{\mathcal{P}}_l\Pi_h\|_{\mathcal{L}(\mathbb{H}_0^1;\mathbb{L}^2)} \leq C\tau\Big[\frac{\alpha}{\sqrt{T-t_l}} + 1\Big] \qquad \forall l = 0, 1, \ldots, N-1\,.$$

Remark 4.4 With the procedure in this section, we can also deduce that, for any $l = 0, 1, \ldots, N-1$,

$$\|\bar{\mathcal{P}}_l\Pi_h - \mathcal{P}_h(t_l)\Pi_h\|_{\mathcal{L}(\dot{\mathbb{H}}_h^2;\mathbb{L}^2)} + \|\widehat{\mathcal{P}}_l\Pi_h - \mathcal{P}_h(t_l)\Pi_h\|_{\mathcal{L}(\dot{\mathbb{H}}_h^2;\mathbb{L}^2)} \leq C\tau\,,$$

which will be applied in Sect. 4.3. We leave the proof to the interested reader.

4.3 Riccati-Based Discretization of Problem (SLQ) and Rates

From Sect. 2.3, we know that the optimal control $U^*(\cdot)$ of Problem **(SLQ)** has a state feedback representation

$$U^*(t) = -\mathcal{P}(t)X^*(t) - \eta(t) \qquad t \in [0, T]\,, \tag{4.98}$$

where $\mathcal{P}(\cdot)$ solves Riccati equation (1.9), and $\eta(\cdot)$ solves a $\mathcal{P}(\cdot)$-dependent PDE (1.10).

In Sect. 4.2, we propose the spatial discretization $\mathcal{P}_h$ solving (4.65), and different spatio-temporal discretization schemes $\mathcal{P}_\cdot$, $\widetilde{\mathcal{P}}_\cdot$, $\bar{\mathcal{P}}_\cdot$ and $\widehat{\mathcal{P}}_\cdot$. Inspired by (4.98) and different schemes of the Riccati equation (4.65), we consider the following *discrete feedback law*

$$U_n = -P_{n+1}X_n - \eta_n \qquad n = 0, 1, \ldots, N-1 \tag{4.99}$$

to discretize Problem **(SLQ)** as follows:

$$\begin{cases} X_{n+1} = A_0X_n + \tau A_0U_n + \big[\beta A_0X_n + A_0\Pi_h\sigma(t_n)\big]\Delta_{n+1}W \\ \qquad\qquad n = 0, 1, \ldots, N-1\,, \\ X_0 = \Pi_h x\,. \end{cases} \tag{4.100}$$

In (4.99), $P_\cdot$ is a spatio-temporal approximation of $\mathcal{P}(\cdot)$, and in this section we choose

$$P_\cdot = \bar{\mathcal{P}}_\cdot \quad \text{or} \quad P_\cdot = \widehat{\mathcal{P}}_\cdot\,;$$

$\eta_\cdot$ is a finite element method-based spatio-temporal discretization scheme of (1.10), which is of the form

$$\begin{cases} \eta_n = A_0\eta_{n+1} + \tau A_0\big[-P_{n+1}\eta_{n+1} + \beta P_{n+1}\Pi_h\sigma(t_{n+1})\big] \quad n = 0, 1, \ldots, N-1\,, \\ \eta_N = 0\,. \end{cases} \tag{4.101}$$

In Chap. 3, we make a spatial discretization scheme of Problem **(SLQ)** and obtain Problem **(SLQ)**$_h$. Similar to Sect. 2.3, Problem **(SLQ)**$_h$ is uniquely solvable, and admits a closed-loop optimal control

$$U_h^*(t) = -\mathcal{P}_h(t)X_h^*(t) - \eta_h(t) \qquad t \in [0, T], \tag{4.102}$$

where $\mathcal{P}_h(\cdot)$ solves (4.65) and $\eta_h(\cdot)$ solves

$$\begin{cases} \eta_h'(t) = -\Delta_h\eta_h(t) + \mathcal{P}_h(t)\eta_h(t) - \beta\mathcal{P}_h(t)\Pi_h\sigma(t) & t \in [0, T], \\ \eta_h(T) = 0. \end{cases} \tag{4.103}$$

The following result is on the regularity of $\eta_h(\cdot)$ in (4.103).

Lemma 4.18 *Suppose that $\sigma(\cdot) \in C\big([0, T]; \mathbb{H}_0^1\big)$, and I_τ is a uniform partition of $[0, T]$ with mesh size $\tau \in (0, 1)$. Then there exists a constant C independent of h such that*

$$\sup_{t\in[0,T]} \|\eta_h(t)\|_{\dot{\mathbb{H}}_h^1}^2 + \int_0^T \|\eta_h(t)\|_{\dot{\mathbb{H}}_h^2}^2\, dt \le C\int_0^T \|\sigma(t)\|^2\, dt, \tag{4.104}$$

$$\max_{0\le k\le N-1} \sup_{t\in[t_k,t_{k+1})} \|\eta_h(t) - \eta_h(t_k)\| \le C\tau\|\sigma(\cdot)\|_{C([0,T];\mathbb{H}_0^1)}. \tag{4.105}$$

Proof (1) Verification of (4.104). By testing (4.103) with $\Delta_h\eta_n(t)$ and applying (4.24) in Lemma 4.2, (3.55) in Lemma 3.6, we can arrive at

$$\begin{aligned} &\frac{1}{2}\|\nabla\eta_h(t)\|^2 + \int_t^T \|\Delta_h\eta_h(s)\|^2\, \mathrm{d}s \\ &= \int_t^T \big(\mathcal{P}_h(s)\eta_h(s) - \beta\mathcal{P}_h(s)\Pi_h\sigma(s), \Delta_h\eta_h(s)\big)_{\mathbb{L}^2}\, \mathrm{d}s \\ &\le \frac{1}{2}\int_t^T \|\Delta_h\eta_h(s)\|^2\, \mathrm{d}s + C\int_t^T \|\eta_h(s)\|^2 + \|\sigma(s)\|^2\, \mathrm{d}s \\ &\le \frac{1}{2}\int_t^T \|\Delta_h\eta_h(s)\|^2\, \mathrm{d}s + C\int_t^T \|\sigma(s)\|^2\, \mathrm{d}s, \end{aligned}$$

which settles the assertion (4.104).

(2) Verification of (4.105). In the same vein as in the proof of (4.79), we can deduce that

$$\begin{aligned} &\|\eta_h(t) - \eta_h(t_k)\| \\ &\le \int_{t_k}^t \big\|E_h(s-t_k)\mathcal{P}_h(s)\big[\eta_h(s) + \beta\Pi_h\sigma(s)\big]\big\|\, \mathrm{d}s \\ &\quad + \int_t^T \big\|\mathcal{P}_h(s)\big[E_h(s-t_k) - E_h(s-t)\big]\big[\eta_h(s) - \beta\Pi_h\sigma(s)\big]\big\|\, \mathrm{d}s \\ &\le C\tau \sup_{t\in[0,T]} \big[\|\eta(t)\| + \|\sigma(t)\|\big] \end{aligned}$$

$$+C\int_t^T \frac{t-t_k}{\sqrt{s-t}} \sup_{s\in[0,T]} \big[\|\eta_h(s)\|_{\dot{\mathbb{H}}_h^1} + \|\Pi_h\sigma(s)\|_{\dot{\mathbb{H}}_h^1}\big]\, \mathrm{d}s$$
$$\le C\tau\|\sigma(\cdot)\|_{C([0,T];\mathbb{H}_0^1)}\,,$$

where C is independent of k. That completes the proof. □

The following is on improved stability bounds for the optimal pair of Problem **(SLQ)**$_h$.

Lemma 4.19 *Suppose that* $\big(X_h^*(\cdot), U_h^*(\cdot)\big)$ *is the optimal pair of Problem* **(SLQ)**$_h$. *Then,* $X_h^*(\cdot), U_h^*(\cdot) \in L^2_{\mathbb{F}}\big(\Omega; C([0,T]; \mathbb{H}_0^1)\big) \cap L^2_{\mathbb{F}}(0,T; \dot{\mathbb{H}}_h^2)$ *and there exists a constant* C *such that*

$$\mathbb{E}\Big[\sup_{t\in[0,T]} \|X_h^*(t)\|^2_{\dot{\mathbb{H}}_h^\gamma} + \sup_{t\in[0,T]} \|U_h^*(t)\|^2_{\dot{\mathbb{H}}_h^\gamma} + \int_0^T \|X_h^*(s)\|^2_{\dot{\mathbb{H}}_h^{\gamma+1}} + \|U_h^*(s)\|^2_{\dot{\mathbb{H}}_h^{\gamma+1}}\, ds\Big]$$
$$\le C\Big[\|x\|^2_{\dot{\mathbb{H}}^\gamma} + \int_0^T \|\sigma(t)\|^2_{\dot{\mathbb{H}}^\gamma}\, dt\Big] \quad \gamma = 0, 1, 2\,. \tag{4.106}$$

Furthermore, if $\beta = 0$, *then for a given uniform partition* I_τ *in time with mesh size* $\tau \in (0, \tau_0] \subset (0, 1)$, *there exists* C *independent of* τ *such that,*

$$\mathbb{E}\Big[\Big\|\int_0^T X_h^*(t) - X_h^*\big(\nu(t)\big)\, dt\Big\|^2\Big]$$
$$\le C\tau^2\Big[\|x\|^2_{\mathbb{H}_0^1} + \sup_{t\in[0,T]} \|\sigma(t)\|^2 + \int_0^T \|\sigma(t)\|^2_{\mathbb{H}_0^1\cap\mathbb{H}^2}\, dt\Big]. \tag{4.107}$$

Proof (1) Verification of (4.106). Partial results have been derived in (3.56a) and (3.56b). We can settle the remaining by relying on the spectral Galerkin method (see e.g. Step (2) in the proof of Theorem 4.5, [14, Sect. 7.1], [15]), $U_h^*(\cdot)$'s feedback representation (4.102), Lemma 4.2, and the BDG inequality.

(2) Verification of (4.107). The proof is similar to that in Lemma 4.8. By equation on $X_h^*(\cdot)$ (SDE (3.35) with $U_h(\cdot) = U_h^*(\cdot)$), we have

$$\mathbb{E}\Big[\Big\|\sum_{n=0}^{N-1}\int_{t_n}^{t_{n+1}} X_h^*(t) - X_h^*(t_n)\, \mathrm{d}t\Big\|^2\Big] = \mathbb{E}\Big[\Big\|\int_0^T X_h^*(t) - X_h^*\big(\nu(t)\big)\, \mathrm{d}t\Big\|^2\Big]$$
$$\le 5\Big\{\mathbb{E}\Big[\Big\|\int_0^T \big[E_h(t) - E_h\big(\nu(t)\big)\big]\Pi_h x\, \mathrm{d}t\Big\|^2\Big]$$
$$+\mathbb{E}\Big[\Big\|\int_0^T\int_0^{\nu(t)} \big[E_h(t-s) - E_h\big(\nu(t)-s\big)\big]U_h^*(s)\, \mathrm{d}s\, \mathrm{d}t\Big\|^2\Big]$$
$$+\mathbb{E}\Big[\Big\|\int_0^T\int_{\nu(t)}^t E_h(t-s)U_h^*(s)\, \mathrm{d}s\, \mathrm{d}t\Big\|^2\Big]$$
$$+\mathbb{E}\Big[\Big\|\int_0^T\int_{\nu(t)}^t E_h(t-s)\Pi_h\sigma(s)\, \mathrm{d}W(s)\, \mathrm{d}t\Big\|^2\Big]$$

$$+\mathbb{E}\Big[\Big\|\int_0^T\int_0^{\nu(t)}\big[E_h(t-s)-E_h\big(\nu(t)-s\big)\big]\Pi_h\sigma(s)\,\mathrm{d}W(s)\,\mathrm{d}t\Big\|^2\Big]\Big\}$$

$$=: 5\sum_{i=1}^{5} J_i\,.$$

Applying Lemma 2.1 and (4.106), we conclude that

$$J_1\le\Big[\int_0^{t_1}2\|x\|\,\mathrm{d}t+C\int_{t_1}^T\tau\nu(t)^{-1/2}\|x\|_{\mathbb{H}_0^1}\,\mathrm{d}t\Big]^2\le C\tau^2\|x\|^2_{\mathbb{H}_0^1}\,,$$

$$\begin{aligned}J_2&\le\mathbb{E}\Big[\int_0^T\int_0^{\nu(t)}\big\|\big[E_h\big(t-\nu(t)\big)-\mathbb{1}_h\big]E_h\big(\nu(t)-s\big)U_h^*(s)\big\|\,\mathrm{d}s\,\mathrm{d}t\Big]^2\\&\le C\,\mathbb{E}\Big[\int_0^T\int_0^{\nu(t)}\tau\big(\nu(t)-s\big)^{-1/2}\sup_{s\in[0,T]}\|U_h^*(s)\|_{\mathbb{H}_0^1}\,\mathrm{d}s\,\mathrm{d}t\Big]^2\\&\le C\tau^2\Big[\|x\|^2_{\mathbb{H}_0^1}+\int_0^T\|\sigma(t)\|^2_{\mathbb{H}_0^1}\,\mathrm{d}t\Big],\end{aligned}$$

and

$$J_3\le C\tau^2\mathbb{E}\Big[\sup_{t\in[0,T]}\|U_h^*(t)\|^2\Big]\le C\tau^2\Big[\|x\|^2+\int_0^T\|\sigma(t)\|^2\,\mathrm{d}t\Big].$$

For J_4, by the mutual independence of $\big\{\int_{t_k}^{t_{k+1}}\int_{t_k}^t E_h(t-\theta)\Pi_h\sigma(\theta)\,\mathrm{d}W(\theta)\,\mathrm{d}t\big\}_{k=0}^{N-1}$, and the Itô isometry, we have

$$J_4\le\tau\sum_{n=0}^{N-1}\int_{t_n}^{t_{n+1}}\mathbb{E}\Big[\int_{t_n}^t\big\|E_h(t-s)\Pi_h\sigma(s)\big\|^2\,\mathrm{d}s\Big]\mathrm{d}t\le C\tau^2\sup_{t\in[0,T]}\|\sigma(t)\|^2\,.$$

For J_5, the Itô isometry and Lemmata 2.1, 2.4 imply

$$\begin{aligned}J_5&\le C\int_0^T\mathbb{E}\Big[\int_0^{\nu(t)}\big\|\big[E_h(t-s)-E_h(t_k-s)\big]\Pi_h\sigma(s)\big\|^2\,\mathrm{d}s\Big]\mathrm{d}t\\&\le C\int_0^T\int_0^{\nu(t)}\tau^2\|\Pi_h\sigma(s)\|^2_{\dot{\mathbb{H}}_h^2}\,\mathrm{d}s\,\mathrm{d}t\\&\le C\tau^2\int_0^T\|\sigma(t)\|^2_{\mathbb{H}_0^1\cap\mathbb{H}^2}\,\mathrm{d}t\,.\end{aligned}$$

A combination of these estimates for J_1 through J_5 then completes the proof. □

For any $l=0,1,\ldots,N-1$, similar to $\mathbb{X}(t_l,T)$, $\mathbb{U}(t_l,T)$ defined in (4.15), we now introduce 'discrete spaces'

$$\mathbb{X}_{\mathbb{A}}\triangleq\Big\{X_\cdot\equiv\{X_n\}_{n=0}^{N-1}\,\Big|\,X_n\in L^2_{\mathcal{F}_{t_n}}(\Omega;\mathbb{V}_h)\quad\forall n=0,1,\ldots,N-1\,,$$

$$\text{and } \tau \sum_{n=0}^{N-1} \mathbb{E}\big[\|X_n\|^2\big] < \infty \Big\},$$

$$\mathbb{U}_{\mathbb{A}} \triangleq \Big\{ U_\cdot \equiv \{U_n\}_{n=0}^{N-1} \,\Big|\, U_n \in L^2_{\mathcal{F}_{t_n}}(\Omega; \mathbb{V}_h) \ \ \forall n = 0, 1, \ldots, N-1,$$

$$\text{and } \tau \sum_{n=0}^{N-1} \mathbb{E}\big[\|U_n\|^2\big] < \infty \Big\},$$

which we endow with the corresponding norms

$$\|X_\cdot\|_{\mathbb{X}_{\mathbb{A}}} \triangleq \Big(\tau \sum_{n=0}^{N-1} \mathbb{E}\big[\|X_n\|^2\big]\Big)^{1/2} \quad \text{and} \quad \|U_\cdot\|_{\mathbb{U}_{\mathbb{A}}} \triangleq \Big(\tau \sum_{n=0}^{N-1} \mathbb{E}\big[\|U_n\|^2\big]\Big)^{1/2}.$$

Similar to Lemma 4.6, we may use (4.101), (4.100) and proceed by induction to have the following representations for $\eta_\cdot$ and $X_\cdot$.

Lemma 4.20 *Suppose that $\sigma(\cdot) \in C([0,T], \mathbb{L}^2)$. Then $\eta_\cdot \in \mathbb{U}_{\mathbb{A}}$, which is of the form*

$$\eta_n = \tau \sum_{k=n+1}^{N} A_0^{k-n}\big[-P_k \eta_k + \beta P_k \Pi_h \sigma(t_k)\big] \quad \forall n = 0, 1, \ldots, N-1. \tag{4.108}$$

For any $U_\cdot \in \mathbb{U}_{\mathbb{A}}$, then $X_\cdot \in \mathbb{X}_{\mathbb{A}}$, and for any $n = 0, 1, \ldots, N$,

$$X_n = A_0^n \Pi_h x + \tau \sum_{k=0}^{n-1} A_0^{n-k} U_k + \sum_{k=0}^{n-1} A_0^{n-k}\big[\Pi_h \sigma(t_k) + \beta X_k\big]\Delta_{k+1} W. \tag{4.109}$$

Proof **(1)** Assertions (4.108) and (4.109) can be derived inductively with the help of (4.101) and (4.100).

(2) Thanks to (4.73) in Lemma 4.12 and (4.96), we know that $\|P_\cdot\|_{\mathcal{L}(\mathbb{L}^2|_{\mathbb{V}_h})}$ is bounded. Then (4.108), together with discrete Gronwall's inequality, leads to

$$\max_{0 \le n \le N} \|\eta_n\| \le C \sup_{t \in [0,T]} \|\sigma(t)\|,$$

by which we have

$$\|\eta_\cdot\|_{\mathbb{U}_{\mathbb{A}}} = \Big(\tau \sum_{n=0}^{N-1} \mathbb{E}\big[\|\eta_n\|^2\big]\Big)^{1/2} \le C \sup_{t \in [0,T]} \|\sigma(t)\|,$$

or equivalently $\eta_\cdot \in \mathbb{U}_{\mathbb{A}}$.

(3) The assertion $X_\cdot \in \mathbb{X}_{\mathbb{A}}$ can be derived by combining (4.109) with the Itô isometry, and applying discrete Gronwall's inequality. □

We are now able to state the main result of this section.

Theorem 4.7 *Suppose that assumption* **(A)** *holds. Let* $\big(X_h^*(\cdot),\,U_h^*(\cdot)\big)$ *be the optimal pair of Problem* $\mathbf{(SLQ)}_h$ *and* $(X_\cdot, U_\cdot)$ *be its approximation given by* (4.99)–(4.101). *Then there exists a constant* C *independent of* h *and* τ *such that*

$$\max_{0\le n\le N}\mathbb{E}\big[\|X_h^*(t_n)-X_n\|^2\big]+\max_{0\le n\le N-1}\mathbb{E}\big[\|U_h^*(t_n)-U_n\|^2\big]\le C\big[\tau^2+|\beta|\tau\big]. \tag{4.110}$$

Proof Firstly, by (4.109), the Itô isometry, feedback form (4.99), discrete Gronwall's inequality and the fact that $\eta_\cdot \in \mathbb{U}_{\mathbb{A}}$, we can deduce that $X_\cdot \in \mathbb{X}_{\mathbb{A}}$. Then we conclude that $U_\cdot \in \mathbb{U}_{\mathbb{A}}$.

The remaining part is on assertion (4.110). Firstly, we prove the error estimate (4.110) for X-part. By (3.35), (4.109) and state feedback representations (4.102), (4.99), we have

$$\begin{aligned}
X_h^*(t_n)-X_n &= \big[E_h(t_n)\Pi_h - A_0^n\Pi_h\big]x \\
&\quad -\sum_{k=0}^{n-1}\int_{t_k}^{t_{k+1}}\Big\{E_h(t_n-t)\big[\mathcal{P}_h(t)-\mathcal{P}_h(t_{k+1})\big]X_h^*(t) \\
&\quad +\big[E_h(t_n-t)-A_0^{n-k}\big]\big[\mathcal{P}_h(t_{k+1})X_h^*(t)+\eta_h(t)\big] \\
&\quad +A_0^{n-k}\big[\mathcal{P}_h(t_{k+1})-P_{k+1}\big]X_h^*(t)+A_0^{n-k}P_{k+1}\big[X_h^*(t)-X_h^*(t_k)\big] \\
&\quad +A_0^{n-k}\big[\eta_h(t)-\eta_h(t_k)\big]+A_0^{n-k}P_{k+1}\big[X_h^*(t_k)-X_k\big] \\
&\quad +A_0^{n-k}\big[\eta_h(t_k)-\eta_k\big]\Big\}\,\mathrm{d}t \\
&\quad +\sum_{k=0}^{n-1}\int_{t_k}^{t_{k+1}}\Big[E_h(t_n-t)\Pi_h\sigma(t)-A_0^{n-k}\Pi_h\sigma(t_k)\Big]\mathrm{d}W(t) \\
&\quad +\beta\sum_{k=0}^{n-1}\int_{t_k}^{t_{k+1}}\Big[E_h(t_n-t)X_h^*(t)-A_0^{n-k}\Pi_hX_k\Big]\mathrm{d}W(t) \\
&=: I_0+\sum_{i=1}^{7}Leb_i+Ito_1+\beta\times Ito_2\,.
\end{aligned} \tag{4.111}$$

Note that Lebesgue integrals on the right-hand side of (4.111) come from continuous and discrete feedback laws. Since feedback laws consist of $\mathcal{P}_h(\cdot)$, $X_h^*(\cdot)$, $\eta_h(\cdot)$ and their approximations $P_\cdot$, $X_\cdot$, $\eta_\cdot$, as well as semigroup $E_h(\cdot)$ and its approximation $E_{h,\tau}(\cdot)$ are involved in the Lebesgue integrals, we use seven terms to distinguish them and each of the Lebesgue integrals addresses specific errors in the scheme: the terms Leb_1, Leb_3 capture the main error effects due to the discretization of the Riccati equation, while the term Leb_2 addresses discretization of $E_h(\cdot)$; the terms Leb_4, Leb_6 describe discretization effects of the optimal state $X_h^*(\cdot)$, while the terms Leb_5, Leb_7 account for discretization effects introduced by (4.101). Finally, the Itô integral Ito_2 is only active in the presence of multiplicative noise when $\beta \neq 0$.

Now we estimate every term on the right-hand sides of (4.111). By (2.10a) in Lemma 2.4,

$$\mathbb{E}\big[\|I_0\|^2\big] \le C\tau^2 \|x\|^2_{\mathbb{H}_0^1 \cap \mathbb{H}^2} \,. \tag{4.112}$$

For Leb_1 which is on the regularity of $\mathcal{P}_h(\cdot)$, by $\mu(\cdot)$ defined in (2.8) and (2.21), we can derive

$$\begin{aligned}
\mathbb{E}\big[\|Leb_1\|^2\big] &\le C\mathbb{E}\Big[\Big\| \int_0^{t_n} E_h(t_n - t)\big[\mathcal{P}_h(t) - \mathcal{P}_h\big(\mu(t)\big)\big] X_h^*(t)\,\mathrm{d}t \Big\|^2\Big] \\
&\le C\mathbb{E}\Big[\Big\| \int_0^{t_n} E_h(t_n - t)\big[E_h\big(2(\mu(t) - t)\big) - \mathbb{1}_h\big] E_h\big(2\big(T - \mu(t)\big)\big) X_h^*(t)\,\mathrm{d}t \Big\|^2 \\
&\quad + \Big\| \int_0^{t_n} E_h(t_n - t) \int_t^{\mu(t)} E_h(s - t)\big(\beta^2 \mathcal{P}_h(s) + \mathbb{1}_h - \mathcal{P}_h^2(s)\big) \\
&\qquad \times E_h(s - t) X_h^*(t)\,\mathrm{d}s\,\mathrm{d}t \Big\|^2 \\
&\quad + \Big\| \int_0^{t_n} E_h(t_n - t) \int_{\mu(t)}^T \big[E_h(s - t) - E_h\big(s - \mu(t)\big)\big] \\
&\qquad \times \big(\beta^2 \mathcal{P}_h(s) + \mathbb{1}_h - \mathcal{P}_h^2(s)\big) E_h(s - t) X_h^*(t)\,\mathrm{d}s\,\mathrm{d}t \Big\|^2 \\
&\quad + \Big\| \int_0^{t_n} E_h(t_n - t) \int_{\mu(t)}^T E_h\big(s - \mu(t)\big)\big(\beta^2 \mathcal{P}_h(s) + \mathbb{1}_h - \mathcal{P}_h^2(s)\big) \\
&\qquad \times \big[E_h(s - t) - E_h\big(s - \mu(t)\big)\big] X_h^*(t)\,\mathrm{d}s\,\mathrm{d}t \Big\|^2\Big] \\
&=: \sum_{i=1}^4 Leb_{1,i} \,.
\end{aligned}$$

For $Leb_{1,1}$, by Lemma 2.4 and (4.106) in Lemma 4.19, it is easy to see

$$Leb_{1,1} \le C\tau^2 \mathbb{E}\Big[\int_0^{t_n} \|X_h^*(t)\|^2_{\mathbb{H}_h^2}\,\mathrm{d}t\Big] \le C\tau^2\Big[\|x\|^2_{\mathbb{H}_0^1} + \int_0^T \|\sigma(t)\|^2_{\mathbb{H}_0^1}\,\mathrm{d}t\Big].$$

By (4.106) in Lemma 4.19, it is easy to check that

$$Leb_{1,2} \le C\tau^2 \sup_{t\in[0,T]} \mathbb{E}\big[\|X_h^*(t)\|^2\big] \le C\tau^2\Big[\|x\|^2 + \int_0^T \|\sigma(t)\|^2\,\mathrm{d}t\Big].$$

For the terms $Leb_{1,3}$ and $Leb_{1,4}$, by applying spectral decomposition, we know that $E_h(t)\mathcal{P}_h(s) = \mathcal{P}_h(s)E_h(t)$ for $t, s \in [0, T]$; then similar to estimate for $Leb_{1,1}$, we have

$$Leb_{1,3} + Leb_{1,4} \le C\tau^2 \mathbb{E}\Big[\int_0^{t_n} \|X_h^*(t)\|^2_{\mathbb{H}_h^2}\,\mathrm{d}t\Big] \le C\tau^2\Big[\|x\|^2_{\mathbb{H}_0^1} + \int_0^T \|\sigma(t)\|^2_{\mathbb{H}_0^1}\,\mathrm{d}t\Big].$$

Now, combining with estimates of $Leb_{1,1}$ through $Leb_{1,4}$, we arrive at

$$\mathbb{E}\big[\|Leb_1\|^2\big] \le C\tau^2\Big[\|x\|_{\mathbb{H}_0^1}^2 + \int_0^T \|\sigma(t)\|_{\mathbb{H}_0^1}^2\,\mathrm{d}t\Big]. \tag{4.113}$$

For Leb_2, Lemma 2.4, (4.104) in Lemma 4.18 and (4.106) in Lemma 4.19 lead to

$$\begin{aligned}\mathbb{E}\big[\|Leb_2\|^2\big] &\le C\,\mathbb{E}\Big[\Big\|\int_0^{t_n} \mathcal{P}_h\big(\mu(t)\big)G_\tau(t_n-t)X_h^*(t)+G_\tau(t_n-t)\eta_h(t)\big]\mathrm{d}t\Big\|^2\Big]\\ &\le C\tau^2\mathbb{E}\Big[\int_0^{t_n} \|X_h^*(t)\|_{\dot{\mathbb{H}}_h^2}^2 + \|\eta_h(t)\|_{\dot{\mathbb{H}}_h^2}^2\,\mathrm{d}t\Big]\\ &\le C\tau^2\Big[\|x\|_{\mathbb{H}_0^1}^2 + \int_0^T \|\sigma(t)\|_{\mathbb{H}_0^1}^2\,\mathrm{d}t\Big].\end{aligned} \tag{4.114}$$

For Leb_3, by Theorem 4.5 (or 4.6) and (4.106) in Lemma 4.19, it follows that

$$\begin{aligned}&\mathbb{E}\big[\|Leb_3\|^2\big]\\ &\le C\,\mathbb{E}\Big[\int_0^{t_n} \big\|\mathcal{P}_h\big(\mu(t)\big)\Pi_h - P_{\pi(t)}\Pi_h\big\|_{\mathcal{L}(\mathbb{H}_0^1;\mathbb{L}^2)} \sup_{t\in[0,T]}\|X^*(t)\|_{\mathbb{H}_0^1}\,\mathrm{d}t\Big]^2\\ &\le C\tau^2\Big[\|x\|_{\mathbb{H}_0^1}^2 + \int_0^T \|\sigma(t)\|_{\mathbb{H}_0^1}^2\,\mathrm{d}t\Big].\end{aligned} \tag{4.115}$$

For Leb_4, by (3.58a) (for general β) or L^2-regularity (4.107), we can deduce that

$$\mathbb{E}\big[\|Leb_4\|^2\big] \le \begin{cases} C\tau\big[\|x\|_{\mathbb{H}_0^1}^2+\|\sigma(\cdot)\|_{C([0,T];\mathbb{L}^2)\cap L^2(0,T;\mathbb{H}_0^1)}^2\big] & \beta\neq 0,\\ C\tau^2\big[\|x\|_{\mathbb{H}_0^1}^2+\|\sigma(\cdot)\|_{C([0,T];\mathbb{L}^2)\cap L^2(0,T;\mathbb{H}_0^1\cap\mathbb{H}^2)}^2\big] & \beta = 0.\end{cases} \tag{4.116}$$

(4.105) yields that

$$\mathbb{E}\big[\|Leb_5\|^2\big] \le C\tau^2\|\sigma(\cdot)\|_{C([0,T];\mathbb{H}_0^1)}^2\,. \tag{4.117}$$

For Leb_6, based on the fact that $\|P_\cdot\|_{\mathcal{L}(\mathbb{L}^2|_{\mathbb{V}_h})}$ is uniformly bounded, it is easy to see that

$$\mathbb{E}\big[\|Leb_6\|^2\big] \le C\tau\sum_{k=0}^{n-1}\mathbb{E}\big[\|X_h^*(t_k) - X_k\|^2\big]. \tag{4.118}$$

For Leb_7, we follow the procedure as in the estimates for Leb_1 through Leb_6 to get

$$\begin{aligned}&\|\eta_h(t_n) - \eta_n\|\\ &\le C\tau\big[\|\sigma(\cdot)\|_{C([0,T];\mathbb{H}_0^1)\cap L^2(0,T;\mathbb{H}_0^1\cap\mathbb{H}^2)} + L_{\sigma,1}\big] + \tau\sum_{k=n+1}^{N}\|\eta_h(t_k) - \eta_k\|,\end{aligned}$$

which, and discrete Gronwall's inequality, lead to

$$\max_{0\leq n\leq N}\|\eta_h(t_n)-\eta_n\|\leq C\tau\big[\|\sigma(\cdot)\|_{C([0,T];\mathbb{H}_0^1)\cap L^2(0,T;\mathbb{H}_0^1\cap\mathbb{H}^2)}+L_{\sigma,1}\big], \tag{4.119}$$

subsequently,

$$\mathbb{E}\big[\|Leb_7\|^2\big]\leq C\tau^2\big[\|\sigma(\cdot)\|^2_{C([0,T];\mathbb{H}_0^1)\cap L^2(0,T;\mathbb{H}_0^1\cap\mathbb{H}^2)}+L^2_{\sigma,1}\big]. \tag{4.120}$$

For Ito_1, we use the Itô isometry, the triangular inequality and Lemma 2.4 to estimate

$$\begin{aligned}\mathbb{E}\big[\|Ito_1\|^2\big]&\leq C\Big[\int_0^{t_n}\big\|E_{h,\tau}(t_n-t)\Pi_h\big[\sigma\big(\nu(t)\big)-\sigma(t)\big]\big\|^2\,\mathrm{d}t\\&\qquad+\int_0^{t_n}\|G_\tau(t_n-t)\sigma(t)\|^2\,\mathrm{d}t\Big]\\&\leq C\int_0^{t_n}\big\|\sigma\big(\nu(t)\big)-\sigma(t)\big\|^2\,\mathrm{d}t+C\int_0^{t_n}\tau^2\|\Pi_h\sigma(t)\|^2_{\dot{\mathbb{H}}_h^2}\,\mathrm{d}t\\&\leq C\tau^2\Big[L^2_{\sigma,1}+\int_0^T\|\sigma(t)\|^2_{\mathbb{H}_0^1\cap\mathbb{H}^2}\,\mathrm{d}t\Big].\end{aligned} \tag{4.121}$$

For Ito_2, by virtue of $\nu(\cdot)$, $\pi(\cdot)$ defined in (2.8) and the Itô isometry, we have

$$\begin{aligned}\mathbb{E}\big[\|Ito_2\|^2\big]&\leq C\mathbb{E}\Big[\int_0^{t_n}\big\|E_{h,\tau}(t_n-t)\big[X_h^*\big(\nu(t)\big)-X_h^*(t)\big]\big\|^2\,\mathrm{d}t\\&\qquad+\int_0^{t_n}\|G_\tau(t_n-t)X_h^*(t)\|^2\,\mathrm{d}t\\&\qquad+\int_0^{t_n}\big\|E_{h,\tau}(t_n-t)\big[X_{\pi(t)-1}-X_h^*\big(\nu(t)\big)\big]\big\|^2\,\mathrm{d}t\Big]\\&=:\sum_{i=1}^3 Ito_{2,i}\,.\end{aligned}$$

By (3.58a), it follows that

$$Ito_{2,1}\leq C\tau\big[\|x\|^2_{\mathbb{H}_0^1}+\|\sigma(\cdot)\|^2_{C([0,T];\mathbb{L}^2)\cap L^2(0,T;\mathbb{H}_0^1)}\big].$$

By (4.106) and Lemma 2.4 with $\gamma=1$, we have

$$Ito_{2,2}\leq C\tau\int_0^{t_n}\mathbb{E}\big[\|X_h^*(t)\|^2_{\mathbb{H}_0^1}\big]\,\mathrm{d}t\leq C\tau\Big[\|x\|^2+\int_0^T\|\sigma(t)\|^2\,\mathrm{d}t\Big].$$

It is easy to see that

$$Ito_{2,3} \leq C\tau \sum_{k=0}^{n-1} \mathbb{E}\big[\|X_h^*(t_k) - X_k\|^2\big].$$

Hence, for Ito_2, we conclude that

$$\mathbb{E}\big[\|Ito_2\|^2\big] \leq C\tau\big[\|x\|^2_{\mathbb{H}_0^1} + \|\sigma(\cdot)\|^2_{C([0,T];\mathbb{L}^2)\cap L^2(0,T;\mathbb{H}_0^1)}\big]$$
$$+C\tau \sum_{k=0}^{n-1} \mathbb{E}\big[\|X_h^*(t_k) - X_k\|^2\big]. \tag{4.122}$$

Finally, combining with (4.111)–(4.122), we have

$$\|X_h^*(t_n) - X_n\|^2_{L^2_{\mathcal{F}_{t_n}}(\Omega;\mathbb{L}^2)} \leq C\tau^2 + C\beta^2\tau + C\sum_{k=0}^{n-1} \tau\|X_h^*(t_k) - X_k\|^2_{L^2_{\mathcal{F}_{t_k}}(\Omega;\mathbb{L}^2)}.$$

Then relying on discrete Gronwall's inequality, we can derive the desired convergence rate for X-part.

Finally, we apply the feedback law to prove the convergence rate for U-part. Relying on (4.99), (4.102), we can see that

$$\begin{aligned} U_h^*(t_n) - U_n &= -\big[\mathcal{P}_h(t_n) - \mathcal{P}_h(t_{n+1})\big]X_h^*(t_n) - \big[\mathcal{P}_h(t_{n+1})\Pi_h - P_{n+1}\Pi_h\big]X_h^*(t_n) \\ &\quad -P_{n+1}\Pi_h\big[X_h^*(t_n) - X_n\big] - \big[\eta_h(t_n) - \eta_n\big] \\ &=: \sum_{i=1}^{4} J_i\,. \end{aligned}$$

With the similar procedure as in the estimate for Leb_1, by applying (4.106), we have

$$\mathbb{E}\big[\|J_1\|^2\big] \leq C\tau^2\Big[\|x\|^2_{\mathbb{H}_0^1\cap\mathbb{H}^2} + \int_0^T \|\sigma(t)\|^2_{\mathbb{H}_0^1\cap\mathbb{H}^2}\,\mathrm{d}t\Big].$$

Remark 4.4 and (4.106) lead to

$$\mathbb{E}\big[\|J_2\|^2\big] \leq C\tau^2\mathbb{E}\big[\|X_h^*(t_n)\|^2_{\dot{\mathbb{H}}_h^2}\big] \leq C\tau^2\Big[\|x\|^2_{\mathbb{H}_0^1\cap\mathbb{H}^2} + \int_0^T \|\sigma(t)\|^2_{\mathbb{H}_0^1\cap\mathbb{H}^2}\,\mathrm{d}t\Big].$$

Convergence rate (4.110) for the X-part and the fact that $\|P_\cdot\|_{\mathcal{L}(\mathbb{L}^2|_{\mathbb{V}_h})}$ is uniformly bounded imply that

$$\mathbb{E}\big[\|J_3\|^2\big] \leq C\,\mathbb{E}\big[\|X_h^*(t_n) - X_n\|^2\big] \leq C[\tau^2 + |\beta|\tau].$$

(4.119) yields that

$$\mathbb{E}\big[\|J_4\|^2\big] \leq C\tau^2.$$

Now, combining with estimates of J_1 through J_4, we derive the rate (4.110) for U-part. That completes the proof. □

Remark 4.5 Relying on Theorem 4.1 and following the procedure in the proof of Theorem 4.7, we can prove that

$$\sup_{t\in[0,T]} \mathbb{E}\big[\|X^*(t) - X_h^*(t)\|^2 + \|U^*(t) - U_h^*(t)\|^2\big] \leq Ch^4 \ln \frac{1}{h}.$$

Remark 4.6 Compared with the open-loop method, the closed-loop method avoids the computation of conditional expectations and the gradient-descent-based iterations (see Algorithm 3.9); hence, it is easier to implement.

The result in Theorem 4.7 shows an improved rate of convergence in case the driving noise in SPDE (1.2) is additive; we do not have a corresponding improvement for the 'open-loop approach' in Chap. 3.

4.4 Numerical Experiments

The scheme in this chapter can be generalized to general SLQ problems; see [5]. Here we only consider the following problem, where the state equation is

$$\begin{cases} \mathrm{d}X(t) = \big[\Delta X(t) + U(t)\big]\mathrm{d}t + \sum_{i=1}^{m} \big[\beta_i X(t) + \sigma_i(t)\big]\mathrm{d}W_i(t) & t \in (0,T], \\ X(0) = x \in \mathbb{H}_0^1 \cap \mathbb{H}^2, \end{cases} \tag{4.123}$$

and a given function $\widetilde{X}(\cdot) : [0,T] \times D \to \mathbb{R}$ is in the cost functional

$$\mathcal{J}\big(U(\cdot)\big) = \frac{1}{2}\mathbb{E}\Big[\int_0^T \|X(t) - \widetilde{X}(t)\|^2 + \|U(t)\|^2 \,\mathrm{d}t + \alpha\|X(T) - \widetilde{X}(T)\|_{\mathbb{L}^2}^2\Big],$$

where $W = (W_1, W_2, \ldots, W_m)$ is a $\mathbb{R}^m$-valued Brownian motion. In this setting, the Riccati equation is

$$\begin{cases} \mathcal{P}'(t) + \Delta\mathcal{P}(t) + \mathcal{P}(t)\Delta + \sum_{i=1}^{m} \beta_i^2 \mathcal{P}(t) + \mathbb{1} - \mathcal{P}^2(t) = 0 & t \in [0,T], \\ \mathcal{P}(T) = \alpha\mathbb{1}, \end{cases} \tag{4.124}$$

and the $\mathcal{P}(\cdot)$-dependent $\eta(\cdot)$ satisfies

$$\begin{cases} \eta'(t) = -\Delta\eta(t) + \mathcal{P}(t)\eta(t) - \sum_{i=1}^{m} \beta_i \mathcal{P}(t)\sigma_i(t) - \widetilde{X}(t) & t \in [0,T], \\ \eta(T) = -\alpha\widetilde{X}(T). \end{cases} \tag{4.125}$$

a. Representation of the problem in vector form. Let $\{\phi_{h,n}\}_{n=1}^{\mathfrak{d}}$ be the finite element basis of $\mathbb{V}_h$ with $\mathfrak{d} \triangleq \dim(\mathbb{V}_h)$. For $1 \le i, j \le \mathfrak{d}$, $1 \le l \le m$, we denote by $M \triangleq (m_{i,j})$, $M_l \triangleq (m^l_{i,j})$ resp. $A \triangleq (a_{i,j})$ mass matrices resp. the stiff matrix, with entries

$$
m_{i,j} = \int_D \phi_{h,j}(x)\phi_{h,i}(x)\,\mathrm{d}x\,, \qquad m^l_{i,j} = \int_D \beta_l(x)\phi_{h,j}(x)\phi_{h,i}(x)\,\mathrm{d}x\,,
$$

$$
\text{resp.} \quad a_{i,j} = \int_D \nabla\phi_{h,j}(x)\nabla\phi_{h,i}(x)\,\mathrm{d}x\,.
$$

We denote by $\mathfrak{X}$ resp. $\mathfrak{U}$ vectors generated by $\mathfrak{x}_n$ resp. $\mathfrak{u}_n$ which satisfy

$$
X_h(\cdot) = \sum_{n=1}^{\mathfrak{d}} \mathfrak{x}_n(\cdot)\phi_{h,n} \quad \text{resp.} \quad U_h(\cdot) = \sum_{n=1}^{\mathfrak{d}} \mathfrak{u}_n(\cdot)\phi_{h,n}\,.
$$

Besides, we denote by $\mathfrak{S}_i(\cdot)$ the vector with coordinates $\big(\mathfrak{s}^i_n(\cdot)\big)$, where $\mathfrak{s}^i_n(\cdot)$ is obtained by $\Pi_h\sigma_i(\cdot) = \sum_{n=1}^{\mathfrak{d}} \mathfrak{s}^i_n(\cdot)\phi_{h,n}$. Similarly, we obtain $\widetilde{\mathfrak{X}}(\cdot)$ related to $\widetilde{X}(\cdot)$.

Then, the vector-valued equation that corresponds to the spatial discretization for (4.123), as well as the matrix-valued equation related to (4.124), and the vector-valued equation related to (4.125) are:

$$
\begin{cases}
M\,\mathrm{d}\mathfrak{X}(t) = \big[-A\mathfrak{X}(t) + M\mathfrak{U}(t)\big]\,\mathrm{d}t \\
\qquad\qquad + \sum_{i=1}^{m} \big[M_i\mathfrak{X}(t) + M\mathfrak{S}_i(t)\big]\,\mathrm{d}W_i(t) \qquad t \in (0,T]\,, \\
\mathfrak{X}(0) = \mathfrak{X}_0\,,
\end{cases}
\tag{4.126}
$$

$$
\begin{cases}
\mathfrak{P}'(t) - \mathfrak{P}(t)M^{-1}A - AM^{-1}\mathfrak{P}(t) + \sum_{i=1}^{m} M_iM^{-1}\mathfrak{P}(t)M^{-1}M_i \\
\qquad\qquad +M - \mathfrak{P}(t)M^{-1}\mathfrak{P}(t) = 0 \qquad t \in [0,T)\,, \\
\mathfrak{P}(T) = \alpha M\,,
\end{cases}
\tag{4.127}
$$

where $\mathfrak{P}(\cdot)$ is the $\mathbb{R}^{\mathfrak{d}\times\mathfrak{d}}$-valued solution, and

$$
\begin{cases}
\mathfrak{F}'(t) + \big[-M^{-1}A - M^{-1}\mathfrak{P}(t)\big]^\top \mathfrak{F}(t) \\
\qquad\qquad + \sum_{i=1}^{m} M_iM^{-1}\mathfrak{P}(t)\mathfrak{S}_i(t) - M\widetilde{\mathfrak{X}}(t) = 0 \qquad t \in [0,T)\,, \\
\mathfrak{F}(T) = -\alpha M\widetilde{\mathfrak{X}}(T)\,.
\end{cases}
\tag{4.128}
$$

For the optimal control $\mathfrak{U}^*(\cdot)$, the feedback law (4.98) takes the form

$$
\mathfrak{U}^*(t) = -M^{-1}\mathfrak{P}(t)\mathfrak{X}(t) - M^{-1}\mathfrak{F}(t) \qquad t \in [0,T]\,.
$$

b. Spatio-temporal discretization. By the second consistent difference Riccati equation and its perturbation shown in Sect. 4.2.3, the corresponding matrix-valued difference Riccati equations are

$$\begin{cases} \bar{\mathfrak{P}}_n = \mathfrak{A}_0^\top \bar{\mathfrak{P}}_{n+1}\mathfrak{A}_0 + \tau \sum_{i=1}^{m} M_i M^{-1}\mathfrak{A}_0^\top \bar{\mathfrak{P}}_{n+1}\mathfrak{A}_0 M^{-1} M_i \\ \qquad -\tau\big[\mathfrak{A}_0^\top \bar{\mathfrak{P}}_{n+1}\mathfrak{A}_0\big]\big[M + \tau \mathfrak{A}_0^\top \bar{\mathfrak{P}}_{n+1}\mathfrak{A}_0\big]^{-1}\big[\mathfrak{A}_0^\top \bar{\mathfrak{P}}_{n+1}\mathfrak{A}_0\big] + \tau M \\ \qquad\qquad n = 0, 1, \ldots, N-1, \\ \bar{\mathfrak{P}}_N = \alpha M, \end{cases} \tag{4.129}$$

and

$$\begin{cases} \widehat{\mathfrak{P}}_n = \big[\mathfrak{A}_0^\top \widehat{\mathfrak{P}}_{n+1}\mathfrak{A}_0\big] + \tau \sum_{i=1}^{m} M_i M^{-1}\big[\mathfrak{A}_0^\top \widehat{\mathfrak{P}}_{n+1}\mathfrak{A}_0\big] M^{-1} M_i \\ \qquad -\tau\big[\mathfrak{A}_0^\top \widehat{\mathfrak{P}}_{n+1}\mathfrak{A}_0\big] M^{-1}\big[\mathfrak{A}_0^\top \widehat{\mathfrak{P}}_{n+1}\mathfrak{A}_0\big] + \tau M \quad n = 0, 1, \ldots, N-1, \\ \widehat{\mathfrak{P}}_N = \alpha M, \end{cases} \tag{4.130}$$

where $\mathfrak{A}_0 = (M + A\tau)^{-1}M$, and $\bar{\mathfrak{P}}_n$, $\widehat{\mathfrak{P}}_n$ are matrices of size $\mathfrak{d} \times \mathfrak{d}$. Here to derive (4.130), we use the following calculation (by setting $V = \mathfrak{A}_0^\top \widehat{\mathfrak{P}}_{n+1}\mathfrak{A}_0$)

$$V(M + \tau V)^{-1}V = VM^{-1}V - \tau V(M + \tau V)^{-1}VM^{-1}V .$$

Now by $\widehat{\mathfrak{P}}_\cdot$, we derive the vector-valued equation of (4.101) as follows

$$\begin{cases} \mathfrak{F}_n = \mathfrak{A}_0^\top \mathfrak{F}_{n+1} - \tau \mathfrak{A}_0^\top \widehat{\mathfrak{P}}_{n+1} M^{-1}\mathfrak{F}_{n+1} + \tau \sum_{i=1}^{m} \mathfrak{A}_0^\top M_i M^{-1}\widehat{\mathfrak{P}}_{n+1}\mathfrak{S}(t_{n+1}) \\ \qquad -\tau \mathfrak{A}_0^\top M \widetilde{\mathfrak{X}}(t_{n+1}) \qquad n = 0, 1, \ldots, N-1, \\ \mathfrak{F}_N = -\alpha M \widetilde{\mathfrak{X}}(T). \end{cases} \tag{4.131}$$

Furthermore, the vector-valued discrete feedback law of (4.99) reads

$$\mathfrak{U}_n = -M^{-1}\widehat{\mathfrak{P}}_{n+1}\mathfrak{X}_n - M^{-1}\mathfrak{F}_n \qquad n = 0, 1, \ldots, N-1, \tag{4.132}$$

such that the discretization of Problem **(SLQ)**, or (4.126) turns to

$$\begin{cases} \mathfrak{X}_{n+1} = \mathfrak{A}_0\mathfrak{X}_n - \tau \mathfrak{A}_0 M^{-1}\big[\widehat{\mathfrak{P}}_{n+1}\mathfrak{X}_n + \mathfrak{F}_n\big] \\ \qquad + \sum_{i=1}^{m} \big[\mathfrak{A}_0 \mathfrak{S}_i(t_n) + \mathfrak{A}_0 M^{-1} M_i \mathfrak{X}_n\big]\Delta_{n+1} W_i \quad n = 0, 1, \ldots, N-1, \\ \mathfrak{X}_0 = \mathfrak{X}_0 . \end{cases} \tag{4.133}$$

c. Numerical experiments. In the following, we present some simulations; these computations were performed on a laptop equipped with an Intel Core i7 (2.7 GHz) and 16 GB of RAM. Also, Fig. 1.1 in Chap. 1 has been obtained by the scheme in part **b**.

Example 4.1 We set the data as follows:

$$D = (0, 1)\,, \quad T = 0.4\,, \quad \alpha = 0\,, \quad m = 1\,, \quad \beta(x) = 3\sin(\pi x)\,,$$
$$\sigma(t, x) = 0\,, \quad \widetilde{X}(t, x) = \frac{2(T - t)}{T}\sin(\pi x)\,, \quad X(0, x) = 2\sin(\pi x)\,.$$

Table 4.1 shows the temporal rates; the operating time is 759 s. To compute rates, we choose $h_{\mathtt{ref}} = \frac{1}{40}$, $\tau_{\mathtt{ref}} = 0.01 \times \frac{1}{2^6}$ to get an approximated optimal pair (the operating time for solving the Riccati equation is 6.2 s) and adopt the Monte-Carlo method to approximate expectations with $\mathtt{M} = 5000$. The left figure in Fig. 4.2 shows the temporal rate of the Riccati-based algorithm.

Table 4.1 Temporal rates of Example 4.1 for different partitions with $h_{\mathtt{ref}} = \frac{1}{40}$, $\tau_{\mathtt{ref}} = 0.01 \times \frac{1}{2^6}$, $\mathtt{M} = 5000$

Time mesh size	Max-$\mathbb{L}^2$-error of state	Rate	Max-$\mathbb{L}^2$-error of control	Rate
0.01×2	1.6235×10^{-1}	$*$	3.2246×10^{-2}	$*$
0.01	1.1493×10^{-1}	0.50	1.6373×10^{-2}	0.98
$0.01 \times \frac{1}{2}$	8.0107×10^{-2}	0.52	8.9435×10^{-3}	0.87
$0.01 \times \frac{1}{2^2}$	5.5334×10^{-2}	0.53	5.1686×10^{-3}	0.79

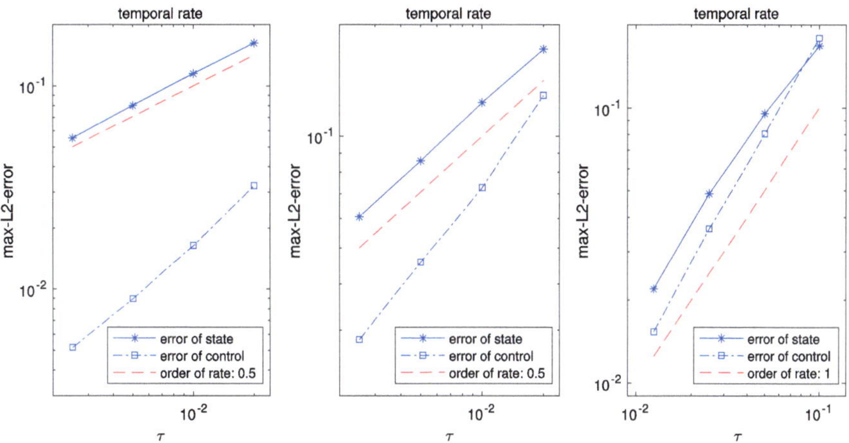

Fig. 4.2 Temporal rates for Examples 4.1, 4.2 and 4.3

Example 4.2 In this example, we consider the case $\alpha > 0$. We set

$$D=(0,1)\,,\quad T=0.26\,,\quad \alpha=1\,,\quad m=3\,,\quad \beta_i(x)=\frac{3}{i}\sin(i\pi x)\,,$$
$$\sigma_i(t,x)=0\,,\quad \widetilde{X}(t,x)=\frac{2(T-t)}{T}\sin(\pi x)\,,\quad X(0,x)=2\sin(\pi x)\,.$$

The simulation is completed in 571 s, and by choosing $h_{\mathtt{ref}}=\frac{1}{40}$, $\tau_{\mathtt{ref}}=0.01\times\frac{1}{2^6}$, the operating time for solving the Riccati equation is 4.6 s. Results are shown in Table 4.2 and Fig. 4.2 (the middle figure).

Both Examples 4.1 and 4.2 show a rate of convergence that is a bit better than $O(\tau^{1/2})$ but clearly stays away from $O(\tau)$—which is in accordance with Theorem 4.7. The next example shows an improvement in order when $\beta=0$; see again Theorem 4.7.

Table 4.2 Temporal rates of Example 4.2 for different partitions with $h_{\mathtt{ref}}=\frac{1}{40}$, $\tau_{\mathtt{ref}}=0.01\times\frac{1}{2^6}$, M=5000

Time mesh size	Max-$\mathbb{L}^2$-error of state	Rate	Max-$\mathbb{L}^2$-error of control	Rate
0.01×2	1.7160×10^{-1}	$*$	1.2889×10^{-1}	$*$
0.01	1.2333×10^{-1}	0.48	7.2609×10^{-2}	0.83
$0.01\times\frac{1}{2}$	8.5754×10^{-2}	0.52	4.5738×10^{-2}	0.67
$0.01\times\frac{1}{2^2}$	6.0641×10^{-2}	0.50	2.8257×10^{-2}	0.69

Example 4.3 Consider the data setting of Example 3.1 in Sect. 3.4.1 with additive noise. The simulation is completed in 1669 s, and by choosing $h_{\mathtt{ref}}=\frac{1}{25}$, $\tau_{\mathtt{ref}}=0.1\times\frac{1}{2^5}$, the operating time for solving the Riccati equation is 0.27 s. Results are shown in Table 4.3 and Fig. 4.2 (the right figure), which clearly show the improved order $O(\tau)$.

Table 4.3 Temporal rates of Example 4.3 for different partitions with $h_{\mathtt{ref}}=\frac{1}{25}$, $\tau_{\mathtt{ref}}=0.1\times\frac{1}{2^5}$, $\mathtt{M}=5000$

Time mesh size	Max-$\mathbb{L}^2$-error of state	Rate	Max-$\mathbb{L}^2$-error of control	Rate
0.1	1.6788×10^{-1}	$*$	1.7890×10^{-1}	$*$
$0.1\times\frac{1}{2}$	9.5150×10^{-2}	0.82	8.0232×10^{-2}	1.16
$0.1\times\frac{1}{2^2}$	4.8618×10^{-2}	0.97	3.6282×10^{-2}	1.14
$0.1\times\frac{1}{2^3}$	2.1887×10^{-2}	1.15	1.5318×10^{-2}	1.24

d. Conclusion. The computational studies in part **c**. evidence superior efficiency of the methods based on the 'closed-loop approach', where the key step is to approximate a discrete version of the stochastic Riccati equation: once this step is taken—which may become more costly for $d=2,3$—, only the simulation of discrete versions of SPDEs is needed; see also Fig. 4.1 in this chapter.

This approach is different from the 'open-loop approach' (see also Fig. 3.1 in Chap. 3), where iterative schemes were set up to compute local discrete gradients, and the simulation of related BSPDEs becomes very costly; see Sect. 3.4. It is, however, that the 'open-loop framework' also allows broader applicability to more general stochastic optimal control problems (including constraints for state and control), while the methods in this chapter efficiently exploit specific properties of Problem **(SLQ)**.

To conclude, we hope that the presented construction of different algorithms in this book, as well as their numerical analysis, and their implementation to compare related complexities will stimulate further research in this field of applied mathematics, leading to efficient algorithms with guaranteed convergence properties for more general stochastic optimal control problems.

References

1. Y. Wang, Error analysis of the feedback controls arising in the stochastic linear quadratic control problems. J. Syst. Sci. Complex. **36**, 1540–1559 (2023)
2. A. Prohl, Y. Wang, Strong rates of convergence for a space-time discretization of the backward stochastic heat equation, and of a linear-quadratic control problem for the stochastic heat equation. ESAIM Control Optim. Calc. Var. **27**, Paper No. 54, 30 (2021)
3. A. Prohl, Y. Wang, Strong error estimates for a space-time discretization of the linear-quadratic control problem with the stochastic heat equation with linear noise. IMA J. Numer. Anal. **42**, 3386–3429 (2022)
4. Q. Lü, P. Wang, Y. Wang, X. Zhang, Chapter 6—numerics for stochastic distributed parameter control systems: a finite transposition method, in *Numerical Control: Part A*, ed. by E. Trélat, E. Zuazua, vol. 23 of Handbook of Numerical Analysis (Elsevier, 2022), pp. 201–232
5. A. Prohl, Y. Wang, Convergence with rates for a Riccati-based discretization of SLQ problems with SPDEs. IMA J. Numer. Anal. **44**, 3393–3434 (2024)
6. J. Zabczyk, *Mathematical Control Theory—an Introduction*, 2nd ed. (Birkhäuser/Springer, Cham, 2020)
7. J. Yong, X.Y. Zhou, Stochastic controls: hamiltonian systems and HJB equations. *Applications of Mathematics (New York)*, vol. 43 (Springer-Verlag, New York, 1999)
8. C. Heij, A.C.M. Ran, F. van Schagen, *Introduction to Mathematical Systems Theory–Discrete Time Linear Systems, Control and Identification* (Birkhäuser/Springer, Cham, 2021)
9. M. Ait Rami, X. Chen, X.Y. Zhou, Discrete-time indefinite LQ control with state and control dependent noises. J. Glob. Optim. **23**, pp. 245–265 (2002)
10. A. MÅ LQVIST, A. PERSSON, T. STILLFJORD, Multiscale differential Riccati equations for linear quadratic regulator problems. SIAM J. Sci. Comput. **40**, A2406–A2426 (2018)
11. T. Stillfjord, Adaptive high-order splitting schemes for large-scale differential Riccati equations. Numer. Algorithms **78**, 1129–1151 (2018)
12. P. Benner, T. Stillfjord, C. Trautwein, A linear implicit Euler method for the finite element discretization of a controlled stochastic heat equation. IMA J. Numer. Anal. **42**, 2118–2150 (2022)
13. M. Kroller, K. Kunisch, Convergence rates for the feedback operators arising in the linear quadratic regulator problem governed by parabolic equations. SIAM J. Numer. Anal. **28**, 1350–1385 (1991)

14. L.C. Evans, Partial differential equations. *Graduate Studies in Mathematics*, vol. 19. (American Mathematical Society, Providence, RI, 1998)
15. Y. Wang, L^2-regularity of solutions to linear backward stochastic heat equations, and a numerical application. J. Math. Anal. Appl. **486**, 123870, 18 (2020)

Index

© The Editor(s) if applicable and The Author(s), under exclusive license to Springer Nature Singapore Pte Ltd. 2026

A. Prohl and Y. Wang, *Numerical Methods for Optimal Control Problems with SPDEs*, SpringerBriefs on PDEs and Data Science,
https://doi.org/10.1007/978-981-95-4469-1

MIX
Papier aus verantwortungsvollen Quellen
Paper from responsible sources
FSC® C105338

If you have any concerns about our products,
you can contact us on
ProductSafety@springernature.com

In case Publisher is established outside the EU,
the EU authorized representative is:
Springer Nature Customer Service Center GmbH
Europaplatz 3, 69115 Heidelberg, Germany

Printed by Libri Plureos GmbH
in Hamburg, Germany